Principles of Biometrical Genetics

The Authors

Dr. Nandakumar Kute (b. 1964) obtained B. Sc. (Ag.) from Dr. PDKV, Akola with distinction, M. Sc. (Ag.) and Ph. D. in Cytogenetics and Plant Breeding from Mahatma Phule Krishi Vidyapeeth, Rahuri with first class and distinction, respectively. He has 25 years of research and teaching experience in various capacities. He is associated in the development of four crop varieties in Mungbean (*Vaibhav*), Pigeonpea (*Vipula*), Chickpea (*Digvijay*)and Grain Amaranthus (*Phule Kartiki*). He guided eight M.Sc. (Ag.) and five Ph.D students. He has taught Principles of Plant Breeding, Principles of Quantitative Genetics, Heterosis Breeding and Improvement of Field Crops for twelve years to PG students. He published fifty three research papers in reputed scientific journals and co-author for one textbook. He serves as external expert for PG programme. He has qualified ASRB NET in Plant Breeding during 2004. At present working as a Associate Professor of Botany at Department of Botany, MPKV, Rahuri, Dist: Ahmednagar(M.S.).

Dr. Gorakshanath Shinde (b. 1963) obtained B.Sc.(Agri.), M.Sc. (Agri.) and Ph.D. in Cytogenetics and Plant Breeding from Mahatma Phule Krishi Vidyapeeth, Rahuri, with distinction. He has 27 years experience of research in Genetics and Plant Breeding. He has been associated with development of 11 crop varieties which include pearl millet, cotton, hybrid napier and marvel grass. He has also been involved in teaching of Ph.D & postgraduate courses *viz.*, Advance Biometrical and Quantitative Genetics and Principles of Quantitative Genetics.. He has published over 40 scientific papers/ articles in national and international journals. Currently, he is working as Sr. Research Assistant at Grass Breeding Scheme, MPKV, Rahuri (MS).

PRINCIPLES OF BIOMETRICAL GENETICS

Dr. Nandakumar Kute
Dr. Gorakshanath Shinde

2016
Daya Publishing House®
A Division of
Astral International Pvt. Ltd.
New Delhi – 110 002

Cataloging in Publication Data--DK
Courtesy: D.K. Agencies (P) Ltd. <docinfo@dkagencies.com>

Principles of biometrical genetics / authors, Dr. Nandakumar Kute, Dr. Gorakshanath Shinde.
pages cm
Includes bibliographical references.

ISBN 978-93-5130-970-3 (International Edition)

1. Biometry. 2. Quantitative genetics--Statistical methods. 3. Population genetics--Statistical methods. I. Kute, Nandakumar, 1964- editor. II. Gorakshanath, Shinde, 1963- editor.

QH323.5.P75 2016 DDC 570.15195 23

Published by : **Daya Publishing House®**
A Division of
Astral International Pvt. Ltd.
– ISO 9001:2008 Certified Company –
4760-61/23, Ansari Road, Darya Ganj
New Delhi-110 002
Ph. 011-43549197, 23278134
E-mail: info@astralint.com
Website: www.astralint.com

Laser Typesetting : **Classic Computer Services**, Delhi - 110 035

Printed at : **Replika Press Pvt. Ltd.**

PREFACE

The ICAR has revised the syllabus and implemented from the year 2009 for M.Sc. (Agri) and it is almost common all over India . So the present book is written for the help of UG and PG students of Agril. Universities. Therefore, an attempt has been made to present the Principles of Quantitative Genetics in a short and very simple language and it will definitely helpful to the UG and PG students of plant breeding in Agriculture Universities of India.

This book almost covers all the topics as per revised ICAR Syllabus for PG and consists of 34 chapters.

A text book hardly claims originality either in content or in presentation style. The present book is no exception and it draws mostly from earlier text books. In addition , We have received help from my colleagues for preparing this quality text book., but the errors, omissions and deficiencies are my own copy write and no one else should be credited for them.

We are also thankful to Shri. P. A. Shinde, University Librarian who insisted an inspired me to write a book and bring my dream in to the truth. Similarly we are also thankful to Astral Book Agency, New Delhi for his all kind of help and timely publishing this book.

We are really grateful to Honorable R. B. Deshmukh, Ex. Vice- Chancellor, MPKV, Rahuri and Hon. Dr.T.A.More Vice- Chancellor MPKV, Rahuri for their moral support and blessings. We also thankful to Dr. B. R. Ulmek, Dean, Faculty of Agril. MPKV, Rahuri, Dr.R. S.Patil , Director of Research , MPKV, Rahuri and Dr. J.V.Patil, Head Department of Botany, MPKV, Rahuri for their inspiration and moral support.

Finally, We record our sincere gratitude to our parents and family members for their patience during the completion of this work.

We appeal to all students and Academicians to send your valuable suggestions and comments so as to further update and improve the quality of this book.

Thanks to one and all.

Dr. Nandakumar Kute

Dr. Gorakshanath Shinde

CONTENTS

1

Introduction

Biometric is the science, which deals with an application of statistical procedures to study the biological problems. A branch of genetics which utilizes various statistical concepts and procedures for genetically studies is called biometrical genetics. There are two branches of biometrical genetics viz., quantitative and population genetics. Quantitative genetics deals with the study of polygenic or quantitative characters and population genetics with frequency of genes in a population.

In quantitative genetics individual genotypes are unidentified, and the traits of individuals are measured. Genotypic values are assigned to genotypes in the population. Quantitative genetics emphasizes the role of selection in controlled populations of known ancestry. A quantitative geneticist observes the phenotype, a product of the genotype and the environment. The genotypic array depends on mating systems and genetic linkage relationships, as well as on allelic frequencies, which in turn are impacted by mutation, migration, random drift, and selection. To make effective observations about phenotypes, the quantitative geneticist has to make assumptions about the mating system, allelic frequency altering forces, and the environment.

COMMON ASSUMPTIONS OF QUANTITATIVE GENETIC ANALYSIS ARE AS FOLLOW:

1. **Reference population defined**: Allelic and genotypic frequencies can only be defined with respect to a specified population. All inferences made about the estimates should depend upon the composition of this reference population.
2. **Absence of linkage**: It is assumed that the trait (phenotype) observed is not affected by autosomal linkage genes.
3. **Presence of diploid Mendelian inheritance**: The plants are assumed to be diploid in which genes segregate and assort independently.
4. **Absence of selection during the formation of inbred lines**: In order for the estimates of genetic variances to pertain to the base reference population, it is required that no selection occur when inbred lines are crossed.
5. **No breeding of the reference population**: It is assumed that the inbreeding coefficient of the reference population is zero. The analysis becomes more complex when inbreeding is coupled with more than two loci and includes the presence of epistasis.

QUANTITATIVE TRAITS

Most traits encountered in plant breeding are quantitatively inherited. Many genes control such traits, each contributing a small effect to the overall phenotypic expression of a trait. Variation in quantitative trait expression is without natural discontinuities (*i.e.,* the variation is continuous). The traits that exhibit continuous variations are also called metric traits. Any attempt to classify such traits into distinct groups is only arbitrary. For example, height is a quantitative trait. If plants are grouped into tall versus short plants, one could find relatively tall plants in the short group and, similarly, short plants in the tall group.

QUALITATIVE GENETICS VERSUS QUANTITATIVE GENETICS

The major ways in which qualitative genetics and quantitative genetics differ may be summarized as:

1. **Nature of traits**: Qualitative genetics is concerned with traits that have Mendelian inheritance and can be described according to kind and can be definitely categorized.

   Quantitative genetics traits are described in terms of the degree of expression of the trait, rather than the kind.
2. **Scale of variability**: Qualitative genetic traits provide discrete (discontinuous) phenotypic variation, whereas quantitative genetic traits produce continuous phenotypic variation.
3. **Number of genes**: In qualitative genetics, the effects of single genes are readily detectable, while in quantitative genetics, single gene effects are not distinct. Rather, traits are under polygenic control (genes with small indistinguishable effects).
4. **Mating pattern**: Qualitative genetics is concerned with individual mating and their progenies. Quantitative genetics is concerned with a population of individuals that may comprise a diversity of mating kinds.
5. **Statistical analysis**: Qualitative genetic analysis is quite straightforward, and is based on counts and ratios. On the other hand, quantitative analysis provides estimates of population parameters (attributes of the population from which the sample was obtained).

THE ENVIRONMENT AND QUANTITATIVE VARIATION

All genes are expressed in an environment (phenotype = genotype + environmental effect). However, quantitative traits tend to be influenced to a greater degree than qualitative traits. It should be pointed out that, under significantly large environmental effects, qualitative traits (controlled by one or a few major genes) can exhibit a quantitative trait inheritance pattern. A strong environmental influence causes the otherwise distinct classes to overlap.

POLYGENIC INHERITANCE

Quantitative traits are controlled by multiple genes or polygenes. Polygenes are genes with effects that are too small to be individually distinguished. They are sometimes called minor genes. In polygenic inheritance, segregation occurs at a large number of loci affecting a trait. The phenotypic expression of polygenic traits is affected due to significant modification by the variation in environmental factors to which plants in the population are subjected. Polygenic variation cannot be classified into discrete groups (*i.e.,* variation is continuous). This is because of the large number of segregating loci, each with effects so small that it is not possible to identify individual gene effects in the segregating population or to meaningfully describe individual genotypes. Instead, biometrics is used to describe the population in terms of means and variances. Continuous variation is caused by environmental variation and genetic variation due to the simultaneous segregation of many genes affecting the trait. These effects convert the intrinsically discrete variation to a continuous one. Biometric genetics is used to distinguish between the two factors that cause continuous variability to occur.

Another aspect of polygenic inheritance is that different combinations of polygenes can produce a particular phenotypic expression. Furthermore, it is difficult to measure the role of the environment on trait expression because it is very difficult to measure the environmental effect on the plant basis. Consequently, a breeder attempting to breed a polygenic trait should evaluate the cultivar in an environment that is similar to that prevailing in the production region. It is beneficial to plant breeding if a tight linkage of polygenes (called polygenic block or linkage block) that has favorable effects on traits of interest to the breeder is discovered.

2

Multiple Factor Hypothesis

In 1910, a Swedish geneticist, Nilsson-Ehle provided a classic demonstration of polygenic inheritance and in the process helped to bridge the gap between our understanding of the essence of quantitative and qualitative traits. Polygenic inheritance may be explained by making three basic assumptions:

1. Many genes determine the quantitative trait.
2. These genes lack dominance.
3. The action of the genes are additive.

In studies on inheritance of seed colour in wheat and oat, he obtained 3:1, 15:1 and 63:1 ratios between coloured and white seeds from different crosses. It is clear from these ratios that the seed colour is governed by one (3:1 ratio in F_2), two (15:1 ratio in F_2) or three (63:1 ratio in F_2) genes. When two or three genes for colour were present together, they seemed to interact in the duplicate manner so that white seed colour was produced only when all genes were present in the recessive state.

However, on a closer examination of the coloured seeds, Nesson-Ehle found that there was a marked difference in the intensity of their colour. Therefore, he further classified the coloured seeds on the basis of intensity of their colour. The red colour from the crosses showing 15:1 ratio could be classified into four distinct classes. (*i*) dark red, (*ii*) medium dark red, (*iii*) medium red (*iv*) light red; they were present in the ratio 1:4:6:4:1 as given below. Thus what was apparently a 15:1 ratio turned out to be a 1:4:6:4:1 ratio on closer examination. This demonstrated for the existence of multiple factors, that is, genes with small and cumulative effects and lacking of dominance.

Parents	$R_1R_1R_2R_2$ Dark Red	×	$r_1r_1r_2r_2$ White
Gametes	R_1R_2		r_1r_2
F_1		$R_1r_1R_2r_2$ Medium Red	

F_2	R_1R_2	R_1r_2	r_1R_2	r_1r_2
R_1R_2	$R_1R_1R_2R_2$ Dark red	$R_1R_1R_2r_2$ Medium dark red	$R_1r_1R_2R_2$ Medium dark red	$R_1r_1R_2r_2$ Medium red
R_1r_2	$R_1R_1R_2r_2$ Medium dark red	$R_1R_1r_2r_2$ Medium red	$R_1r_1R_2r_2$ Medium red	$R_1r_1r_2r_2$ Light red
r_1R_2	$R_1r_1R_2R_2$ Medium dark red	$R_1r_1R_2r_2$ Medium red	$r_1r_1R_2R_2$ Medium red	$r_1r_1R_2r_2$ Light red
r_1r_2	$R_1r_1R_2r_2$ Medium red	$R_1r_1r_2r_2$ Light red	$r_1r_1R_2r_2$ Light red	$r_1r_1r_2r_2$ White

F_2 Phenotypic Ratio: 1 Dark Red: 4 Medium dark red: 6 Medium red: 4 Light red: 1 White

According to this hypothesis, inheritance of seed colour in crosses showing 1:4:6:4:1 ratio in F_2 may be explained as follows. The seed colour in such crosses is governed by two genes, Say R1 and R2 (Positive alleles), r_1 and r_2 being the negative alleles of these genes. If the two parents of such across have the phenotypes dark red and white, they would have the genotypes $R_1R_1R_2R_2$ and $r_1r_1r_2r_2$, respectively. These parents will produce the gametes R_1R_2 and r_1r_2, respectively which will unite to give rise to the $R_1r_1R_2r_2$ genotype in F_1. Since the F_1 seeds have only two positive alleles, their colour will be intermediate between those of two parents *i.e.,* medium red. The following is the genotypic and phenotypic frequencies produced by the segregation of two genes of R_1 and R_2 with cumulative effect on seed colour in wheat.

Genotype	Frequency	No. of dominant alleles	Phenotype	Frequency
$R_1R_1R_2R_2$	1	4	Dark red	1
$R_1r_1R_2R_2$	2	3	Medium dark red	4
$R_1R_1R_2r_2$	2	3	Medium dark red	
$R_1r_1R_2r_2$	4	2	Medium red	6
$R_1R_1r_2r_2$	1	2	Medium red	
$r_1r_1R_2R_2$	1	2	Medium red	
$R_1r_1r_2r_2$	2	1	Light red	4
$r_1r_1R_2r_2$	2	1	Light red	
$r_1r_1r_2r_2$	1	0	White	1

From the above data, the intensity of seed colour depends on the number of positive alleles of all the genes affecting seed colour present in a seed and not on which of these genes is present in the positive or the negative form. In other words, a quantitative character is governed by several genes; the positive alleles of all the genes governing the trait are similar to each other in their action (effect), and that their effects are additive or cumulative in nature. This is the essence of the multiple factor inheritance.

Other difficulties associated with studying the genetics of quantitative traits are dominance, environmental variation, and epistasis. Not only can dominance obscure the true genotype, but both the amount and direction can vary from one gene to another. For example, allele *A* may be dominant to *a*, but *b* may be dominant to *B*. It has previously been mentioned that environmental effects can significantly obscure genetic effects. Non-allelic interaction is a clear possibility when many genes are acting together.

As the number of genes controlling a trait increases, the phenotypic classes become increasingly indistinguishable. Given n genes, the number of possible phenotypes in the F_2 is given by $2n + 1$

Number of gene loci	1	2	3	-----	n
Ratio of F_2 individuals expressing either extreme Phenotype (parental)	1/4	1/16	1/64	-----	$(1/4)^n$

3 Biometrical Techniques in Plant Breeding

ANALYSIS OF POLYGENIC TRAITS

Polygenic traits require metric measurements and analysis of such traits are based on means, variance and covariances. The various statistical procedures which are used for analysis of polygenic traits are:

1. Simple measures of variability
2. Tests of significance
3. Analysis of variance
4. Analysis of covariance
5. Regression analysis
6. Correlation analysis

BIOMETRICAL TECHNIQUES

Various statistical procedures which are employed in biometrical genetics are called as biometrical techniques. In plant breeding research, various biometrical techniques are used for improvement of yield and its contributing traits. On the basis of their application in plant breeding, biometric techniques can be divided into four major groups.

1. Assessment of variability
2. Selection of elite genotypes
3. Choice of suitable parents and breeding procedures
4. Assessment of stability of genotypes.

Assessment of variability: The success of genetic improvement in any characters depends on nature of variability present in the gene pool of the character. Hence, assessment of existing variability for any character present in the gene pool of a crop species is of utmost importance to the breeder for starting a judicious plant breeding programme. Biometrical techniques for systemic assessment of variability are

(*i*) *Simple measures of variability* : These include range, mean, standard deviation, variance, standard error, coefficient of variation, covariance etc.

(*ii*) *Components of variation* : These include phenotypic, genotypic and environmental variation, heritability and genetic advance etc.

(*iii*) Metroglyph analysis (*iv*) D^2 analysis

(*v*) Principal component analysis (*vi*) Molecular diversity analysis

Selection of elite genotypes: The selection of the superior genotypes from the breeding population is the most important activity in all plant breeding programmes. The biometrical techniques viz., Correlation analysis, path analysis and discriminant function analysis provide information about the relative contribution of component traits toward yield and help in the selection of superior genotypes from the breeding population.

Choice of suitable parents and breeding procedures: The selection of suitable parents for hybridization is one of the most important steps in the breeding programme. It is essential that parents should be choosen on the basis of their genetic value. Normally, combining ability effects prove the genetic value of the parents and hybrids, therefore parents are chosen based on their combining ability. After selection of parents, the breeding procedure to be followed for successful results mostly depends upon the genetics of the quantitative character. Therefore, before starting any breeding programme, it is essential to study the gene action involved in the expression of various quantitative traits. The biometrical techniques which are used for study of gene action are, namely, combining ability analysis (Line × tester analysis, diallel, partial diallel, triallel and quadriallel), generation mean analysis, biparental cross analysis and triple test cross analysis.

List of various Biometrical Techniques used in Crop Improvement

S.No.	Application	Biometrical techniques used	Genetic information obtained
1.	**Assessment of variation**		
		(*i*) Measures of dispersion : Range, Standard deviation, Variance, Standard Error, Coefficient of Variation	Phenotypic variabilty
		(*ii*) Components of genetic variances: Phenotypic, Genotypic and Environmental variances & PCV, GCV, ECV,	heritability, genetic advance, gene action
		(*iii*) Metroglyph Analysis	Variability and diversity
		(*iv*) D^2 statistics	Genetic diversity
		(*v*) Principal component analysis	Genetic diversity
2.	**Selection of elite genotypes**		
		(*i*) Correlation analysis	Character association
		(*ii*) Path coefficient analysis	Cause and effects
		(*iii*) Discriminant function analysis	Selection criteria

Contd...

Contd...

3	**Choice of parents and breeding procedures**		
	(*a*) Analysis of several single crosses		
		(*i*) Diallel cross analysis	GCA, SCA variances and effects and additive & dominance components
		(*ii*) Partial diallel analysis	GCA effects and variances and additive & dominance components
		(*iii*) Line × tester Analysis	GCA, SCA variances and effects and additive & dominance components
	(*b*) Analysis of three way and double crosses		
		Triallel & quadriallel analysis	Additive , dominance & epistasis variances
	(*c*) Analysis of individual cross		
		(*i*) Generation mean analysis	Additive, dominance and epistasis variances
		(*ii*) Biparental cross analysis	Additive and dominance variance
		(*iii*) Triple test cross analysis	Additive, dominance variances and presence/absence of epistasis
4	**Assessment of Varietal Adaptability**		
		Stability analysis models	Phenotypic stability

Assessment of stability of the genotypes: Stability analysis is useful in the identification of adaptable geneotypes and in predicting the response of various genotypes over changing environments. There are different stability models *viz.*, Parametric model , regression model , non parametric and multivariate analysis (AMMI, cluster analysis, principal component analysis, pattern analysis etc.).

4 Analysis of Continuous Variation

For any crop improvement programme, variability is a must in the base population which should be periodically assessed. In most of the plant breeding experiments enormous data is being recorded on various aspects of plant population to study the extent of variability. So an elementary knowledge of statistics is essential for plant breeders and geneticists for the genetic interpretation of results. Continuous variation present in the population is partly heritable and partly non heritable. The non heritable components arise from the influence of external environment, where as heritable component depend on genes at many loci working together. The continuous variation can be analyzed only by estimating the biometrical qualities like arithmetic mean, mode, median, range, standard deviation, standard error, coefficient of variation, variances and covariances.

ESTIMATION OF VARIABILITY PARAMETERS

Simple measures

Mean of the distribution (A first degree statistic)

(*i*) **Arithmetic mean ($\bar{x}$)**: It is defined as the sum of all observations in a sample divided by their number (or) it is the average value of a set of observations (or) a data.

$$\bar{x} = \frac{\Sigma x}{N}$$

Where,

Σ = summation

x = an observation

N = number of observations

(*ii*) **Mode:** Most frequent values in a population.

(*iii*) **Median:** Middle value in an array of observations.

(*iv*) **Geometric Mean:** It is a midpoint on multiplying geometric scale. The geometric mean of two numbers is the square root of their product. *e.g.* It is the term between 2 and 8 in geometric series (2,4,8).

Frequency distribution: Sometimes the raw data is grouped into classes and the frequency in each class is recorded. The data is recorded as frequency distribution and the mean is calculated as.

$$\text{Mean} = \frac{\Sigma fx}{N}$$

Where f = class frequency, x = class value and N= the number of observations

MEASURES OF DISPERSION (SECOND DEGREE STATISTICS)

The degree to which numerical data tend to spread about the mean value is called dispersion. The following are the measures for measurement of dispersion of individual observations in the population.

(*i*) **Range:** Range is the difference between the lowest and the highest value present in the observation.

(*ii*) **Mean Deviation or Average Deviation:** It is the average deviation of individual observations from the mean

$$\text{Mean Deviation} = \sqrt{\frac{\Sigma(x-\bar{x})}{N}}$$

Where,

Σ = summation

x = an observation

N = number of observations

$$\text{Mean deviation by frequencies} = \sqrt{\frac{\Sigma fd}{N}}$$

Where f = class frequency, d = deviation of class value from mean and N = the number of observations

(*iii*) **Standard deviation:** It is the square root of arithmetic mean of squares of all the deviations measured from the mean. It is also the square root of variances and calculated using the formula

$$SD = \sqrt{\frac{\Sigma x^2 - \frac{(\Sigma x)^2}{N}}{N-1}}$$

(*iv*) **Variance:** It is defined as the average of squared deviation of all the individual observations from the mean (or) it is the difference between the sums of squares of all the observations in a sample from its mean divided by the degrees of freedom.

$$\text{Variance} = \frac{\Sigma x^2 - \frac{(\Sigma x)^2}{N}}{N-1}$$

(*v*) **Standard error:** It is defined as the measure of difference between the sample mean (and the population mean). It is a measure of uncontrolled variation present in a sample. It is the ratio of standard deviation to the square root of number of observations. It is customary to denote any sample in terms of its mean ± SE

$$SE = \frac{SD}{\sqrt{N}}$$

(*vi*) **Coefficient of variation:** All the above measures of variation are absolute being expressed in terms of units in which they are recorded. A measure of variation which is independent of the unit of measurement is coefficient of variation. It is the ratio of standard deviation of a sample to its mean expressed as percentage

$$CV\% \quad = \quad \frac{SD}{Mean} \times 100$$

(*vii*) **Covariance**: Any pair of related characters may yield a covariance.

3. **Higher order degree statistics:** In biometrical genetics, this statistics is used to study the mean skewness of F_2, F_3 etc.

SOLVED EXAMPLE

Work out the biometrical quantities from the following data measured for crude protein content on 24 genotypes of pearl millet .

Genotypes	Crude protein (%)	Genotypes	Crude protein (%)	Genotypes	Crude protein (%)
1	4.8	9	5.2	17	7.4
2	6.6	10	6.3	18	8.3
3	5.7	11	8.9	19	5.3
4	4.7	12	8.7	20	5.5
5	6.1	13	7.2	21	5.5
6	7.8	14	11.5	22	4.9
7	5.9	15	12.1	23	9.1
8	12.6	16	9.2	24	13.0
				Total	**182.3**

(*a*) **Arithmetic mean** $= \frac{\Sigma x}{N}$

$$= \frac{4.8 + 6.6 + 13.0}{24} = \frac{182.3}{24}$$

$= \mathbf{7.6}$

(*b*) **Range** $= 4.7 \text{ to } 13.0 = 13.0 - 4.7 = 8.3$

(*c*) **Variance** $= \dfrac{\Sigma x^2 - \dfrac{(\Sigma x)^2}{N}}{N-1}$

$$= \frac{4.8^2 + 6.6^2 + ... + 13.0^2 - \dfrac{(182.3)^2}{24}}{23}$$

$$= \frac{1537.63 - 1384.72}{23} = 6.65$$

(*d*) **Standard deviation** $= \sqrt{\text{Variable}}$

$$= \sqrt{\sigma^2 p} = \mathbf{2.58}$$

(*e*) **Standard error** $= \frac{\text{SD}}{\sqrt{\text{N}}} = \frac{2.58}{4.90} = \mathbf{0.53}$

(*f*) **Coefficient of variation (%)** $= \frac{\text{SD}}{\text{Mean}} \times 100 = \frac{2.58}{7.60} \times 100 = \mathbf{33.95}$

CONCLUSION

From the above estimates, it is revealed that the crude protein (%) ranged from 4.7 (Genotype 4) to 13.0 (Genotype 24) having general mean (7.6 %). The estimates of variance, standard deviation and C.V. % were also high in magnitude which indicated high variability among the 24 genotypes of pearl millet for crude protein content.

5

Analysis of Variance

The variability present in a population is of polygenic nature and this polygenic variation is of three types *viz.*, Phenotypic, Genotypic and Environmental.

The statistical procedure which separates (or) splits the total variation into different components is called analysis of variance (or) ANOVA. It is useful in estimating the different components of variance. It provides basis for the test of significance and it is carried out only with replicated data obtained from standard statistical experimental results.

STEPS INVOLVED IN ANOVA

1. First total of all the observations (Treatment wise, Replication wise and Grand Total)
2. Find out the (AM) = Grand Total/N
3. Find out correction factor(CF) = $[GT]^2/N$
4. Find the total sum of squares(TSS) = Square all individual observations and add & minus CF
5. Treatment sum of squares (T*rss*) = Add the squares of individual treatment totals and divide it by number of replications minus CF
6. Replication sum of squares (RSS) = Add the squares of individual replication totals and divide it by number of treatments minus CF
7. Error sum of squares (ESS) = TSS – (T*r*SS+ RSS)

Analysis of Variance for RBD

Source	*df*	Sum of squares	Mean squares	F value
Replications	$(r-1)$	RSS	$MSr = \frac{RSS}{r-1}$	MSr/MSe
Treatments	$(t-1)$	T*r*SS	$MSt = \frac{Trss}{t-1}$	MSt / MSe
Error	$(r-1)(t-1)$	ESS	$MSe = \frac{ESS}{(r-1)(t-1)}$	
Total	$(rt-1)$	TSS		

F (calculated) is compared with F (Table) value by looking at the F table for treatments *df* $(t-1)$ and error *df* values $(r-1)(t-1)$. If the calculated F value is greater than F (Table value) then it is significant.

From the above table, environmental, genotypic and phenotypic variances are estimated as suggested by Lush (1940).

Environmental Variance : It is the non heritable variation which is due to the environment which varies depending upon the environments. It is denoted as VE or σ^2_e and calculated as

$$VE\ (\sigma^2_e) = \quad MSe$$

Genotypic Variance: It is the inherent variation which remains unaltered by the environment. It is the variation due to genotypes. It is denoted by VG or σ^2_g and is calculated using the formula

$$\text{Genotypic Variance } (\sigma^2_g) = \frac{MSt - MSe}{r}$$

Where, MSt = mean sum of squares of treatments

MSe = mean sum of squares of error

r = number of replications

Phenotypic Variance: It is the total variation which is observable and is the sum total of genotypic and environmental variances. It is represented by VP or σ^2_p and calculated as

$$\text{Phenotypic Variance } (\sigma^2_p) = 6^2_g + 6^2_e$$

ANOVA is also useful in the estimation of phenotypic, genotypic and environmental coefficients of variation (Burton, 1952).

Phenotypic coefficient of variation (PCV): It is defined as the ratio of phenotypic standard deviation to the mean expressed as percentage and is calculated using the formula

$$PCV = \frac{\sqrt{\sigma^2 p}}{\text{Mean}} \times 100 \text{ or } \frac{\sigma p}{\text{Mean}} \times 100$$

Genotypic coefficient of variation (GCV): It is the ratio of genotypic standard deviation to the mean expressed as percentage and is calculated using the formula

$$GCV = \frac{\sqrt{\sigma^2 g}}{\text{Mean}} \times 100 \text{ or } \frac{\sigma p}{\text{Mean}} \times 100$$

Environmental coefficient of variation (ECV): It is the ratio of environmental standard deviation to the mean expressed as percentage and is calculated using the formula

$$ECV = \frac{\sqrt{\sigma^2 e}}{\text{Mean}} \times 100 \text{ or } \frac{\sigma p}{\text{Mean}} \times 100$$

GCV and PCV are classified as suggested by Sivasubramoniam and Madhavamenon in 1973 as Low (less than 10 %) ; Moderate (10-20 %) and High (more than 20%).

If the difference between GCV and PCV is less, it indicates that environment has little influence on the characters and if the difference is more it states that environment plays a substantial role in the expression of that character.

SOLVED EXERCISE

Twelve genotypes of pearl millet were evaluated in RBD with three replications. Green forage yield of five competitive plants in each replication was measured and means were calculated. Mean data was analyzed as per RBD design.

Sr.No.	Genotype	Green forage yield /plant (kg)				
		R-I	R-II	R-III	Total	Mean
1	G–1	3.59	3.06	2.90	9.55	3.18
2	G-2	2.67	3.62	3.00	9.29	3.10
3	G-3	3.29	2.45	2.60	8.34	2.78
4	G-4	3.92	3.33	3.55	10.80	3.60
5	G-5	3.12	2.50	2.70	8.32	2.77
6	G-6	4.51	3.50	3.82	11.83	3.94
7	G-7	5.01	6.33	5.50	16.84	5.61
8	G-8	5.68	6.90	6.30	18.88	6.29
9	G-9	3.40	3.88	3.55	10.83	3.61
10	G–10	3.74	3.92	3.35	11.01	3.67
11	G–11	3.95	4.34	4.10	12.39	4.13
12	G–12	4.88	4.21	4.40	13.49	4.50
	Total (R)	47.76	48.04	45.77	141.57	3.93

$$\text{Correction Factor} = \frac{(GT)^2}{(r \times t)} = \frac{(141.57)^2}{(3 \times 12)} = \frac{20042.06}{36} = 556.72$$

$$\text{Total SS} = \sum x^2 - CF = (3.59)^2 + ... + (4.40)^2 - 556.72 = 599.91 - 556.72 = 43.19$$

$$\text{Treat.SS} = \frac{\sum T^2}{r} = CF = (9.55)^2 + ... + \frac{(13.49)^2}{3} - 556.72 = \frac{1782.92}{3} - 556.72 = \mathbf{38.91}$$

$$\text{Repl.SS} = \frac{\sum T^2}{T} - CF = (47.76)^2 + ... + \frac{(45.77)^2}{12} - 556.72 = \frac{6683.75}{12} - 556.72 = \mathbf{0.26}$$

$$\text{Error SS} = \text{Total S.S} - \text{Treat.SS} - \text{Repl.SS} = 43.20 - 38.92 - 0.26 = \mathbf{4.02}$$

Analysis of variance for green forage yield (kg/plant)

Source	***df***	**Sum of squares**	**Mean sum squares**	**F cal**	**F table 5 %**
Repl.	$(r - 1) = 2$	0.26	0.13	0.70	3.44
Genotypes	$(t - 1) = 11$	38.91	3.54*	19.34	2.26
Error	$(r - 1)(t - 1) = 22$	4.02	0.18		
Total	$(r \times t) - 1 = 35$	43.19			

$$\text{SE} = \sqrt{\frac{\text{EMS}}{2}}\sqrt{\frac{0.18}{3}} = 0.25$$

$$\text{CD at 5\%} = \text{SE} \times \sqrt{2}\ t \text{ value at error } df$$

$$= 0.25 \times 1.414 \times 2.07 = \mathbf{0.73}$$

$$\text{CV\%} = \frac{\sqrt{\text{EMS}}}{\text{Mean}} \times 100\ \frac{\sqrt{0.18}}{3.93} \times 100 = \mathbf{10.68}$$

CONCLUSION

From the analysis of variance, the F ratio was highly significant indicating that there were significant differences among the 12 genotypes of pearl millet for green forage yield. Among the genotypes, G-8 and G-7 recorded significantly highest green forage yield than rest of the genotypes, while G-7 was on par with G-8.

ESTIMATION OF PHENOTYPIC, GENOTYPIC AND ENVIRONMENTAL COMPONENTS OF VARIANCES FROM RBD ANOVA

$$\text{Genotypic variance } (\sigma^2_g) = \frac{\text{MSt} - \text{MSe}}{r} = \frac{3.54 - 0.18}{3} = \frac{3.36}{3} = 1.12$$

$$\text{Environmental variance } (\sigma^2_e) = \text{MSe} = 0.18$$

$$\text{Phenotypic variance } (\sigma^2_p) = \sigma^2_g + \sigma^2_e = 1.12 + 0.18 = 1.30$$

ESTIMATION OF PHENOTYPIC, GENOTYPIC AND ENVIRONMENTAL COEFFICIENTS OF VARIABILITY FOR POLYGENIC TRAIT

Phenotypic Coefficient of Variation (PCV %)

$$= \frac{\sigma p}{\text{Mean}} \times 100 = \frac{\sqrt{1.30}}{3.93} \times 100 = \frac{1.14}{3.93} \times 100 = 29.00$$

Genotypic Coefficient of Variation (GCV %)

$$= \frac{\sigma g}{\text{Mean}} \times 100 = \frac{\sqrt{1.12}}{3.93} \times 100 = \frac{1.06}{3.93} \times 100 = 26.97$$

Environmental Coefficient of Variation (ECV %)

$$= \frac{\sigma e}{\text{Mean}} \times 100 = \frac{\sqrt{0.18}}{3.93} \times 100 = \frac{0.42}{3.93} \times 100 = 10.69$$

CONCLUSION

In above study, moderate ECV and high GCV and PCV estimates for green forage yield were observed indicating high variability among the twelve pearl millet genotypes for green forage yield.

6 Analysis of Covariance

A statistical procedure which simultaneously splits the variation present in two characters simultaneously into different components such as variation due to genotypes, variation due to phenotypes and variation due to error is called as analysis of co variance or ANACOVA. It is carried out with replicated data obtained from any standard statistical design. ANOVA and ANACOVA have several joint applications in plant breeding. *i.e.*, for estimation of phenotypic, genotypic and environmental correlations and path coefficients, D^2 statistics, selection indices and regression coefficients.

Analysis of Covariance for RBD

Source	*df*	Sum of product	Mean products	F value
Replication	$(r-1)$	RSP	MSPr = RSP/ $r-1$	MSPr/MSPe
Treatment	$(t-1)$	TrSP	MSPt = TrSP/ $t-1$	MSPt/MSPe
Error	$(r-1)(t-1)$	ESP	MSPe = ESP/ $(r-1)(t-1)$	
Total	$(rt-1)$	TSP		

Genotypic covariance: It is the genetic variation due to genotype when two characters are simultaneously studied. It is represented as G.COV and is calculated as

$$\text{G.COV} = \frac{\text{MSP}t - \text{MSP}e}{r}$$

MSPt = mean sum of products for treatments

MSPe = mean sum of products for error

r = number of replications

Environmental covariance: It is the non heritable variation due to error of the environment. It is represented as ECOV and calculated as ECOV = MSPe

Phenotypic covariance: It is the sum of genotypic covariance and environmental covariance.

$$\text{PCOV} = \text{GCOV} + \text{ECOV}$$

SOLVED EXERCISE

Twelve varieties of sesame were evaluated in RBD with three replications. The following are the data recorded for number of capsules (x) and seed yield (y). From the data estimate the components of co variance.

Variety	(x) No. of capsules				(y) Seed yield (g)			
	R-I	R-II	R-III	Total	R-I	R-II	R-III	Total
V–1	114	109	98	321	47	39	39	125
V-2	62	71	70	203	32	36	37	105
V-3	72	70	75	217	24	18	23	65
V-4	50	55	57	162	22	26	24	72
V-5	89	94	89	272	35	35	32	102
V-6	90	85	87	262	34	33	29	96
V-7	83	81	76	240	33	30	28	91
V-8	73	69	68	210	28	26	25	79
V-9	89	85	85	259	37	37	33	107
V–10	97	101	103	301	34	30	31	95
V–11	54	51	54	159	26	30	29	85
V–12	57	55	51	163	20	23	20	63
Total	930	926	913	2769	372	363	350	1085

$$\text{Correction factor (CF)} = \frac{\text{GT}x - \text{GT}y}{r \times t} = \frac{2769 \times 1085}{(3 \times 12)} = \frac{3004365}{36} = 83454.58$$

$$\text{Total SP} = \sum(xy) - \text{CF} = (114 \times 47) + ... + (51 \times 20) - 83454.58$$

$$= 86337.0\text{-}83454.58 = 2882.42$$

$$\text{Treat SP} = \frac{\Sigma \text{T}x\text{T}y}{r} - \text{CF}$$

$$= \frac{(321 \times 125) + ... + (163 \times 63)}{3} - 83454.58 = 2754.42$$

$$\text{Repl.SP} = \frac{\Sigma \text{R}x\text{R}y}{t} - \text{CF} = \frac{1001648}{312} - 83454.58$$

$$= 83470.67\text{-}83854.58 = 16.08$$

$$\text{Error SP} = \text{Total SP} - \text{Treat.SP} - \text{Repl.SP}$$

$$= 2885.42 - 2754.42 - 16.08 = 111.92$$

Analysis of covariance for RBD

Source	*df*	SP	MSP	F cal
Replications	$(r-1)=2$	16.08	8.04	1.58
Genotypes	$(t-1)=11$	2754.42	250.40**	49.22
Error	$(r-1)(t-1)=22$	11.92	5.09	
Total	$(r \times t - 1)=35$	2882.42		

$$\text{Environmental covariance} = \sigma e_{xy} = [\text{Error MP}] = 5.09$$

$$\text{Genotypic covariance} = \sigma g_{xy} = \frac{[\text{MP genotypes} - \text{MP error}]}{r}$$

$$= \frac{250.40 - 5.09}{3} = 81.77$$

$$\text{Phenotypic covariance} = \sigma p_{xy} = [\sigma g_{xy} + \sigma e_{xy}] = 81.77 + 5.09 = 86.86$$

$$\text{Coheritability } (xy) = \frac{\text{Cov. genotypic } (\sigma\, gxy)}{\text{Cov. Phenotypic } (\sigma\, pxy)} \times 100$$

$$= \frac{81.77}{86.86} \times 100 = 91.14\%$$

CONCLUSION

The F ratio was highly significant indicating that there is significant differences occurring simultaneously in two characters studied and also high co-heritability between these two characters.

7

Variation Associated With Polygenic Traits

The variation in quantitative characters arises from the joint action of the genotype and environment. This joint action can be expressed in a linear fashion.

$$A = \mu + g + e + (ge)$$

Where,

A = the phenotypic expression of the character

μ = general population mean

g = genotypic effect

e = an environmental effect

ge = an interaction effect

These above components of above variation associated with polygenic traits can be share out and estimated with the use of different experimental populations.

1. **Estimation of Environmental Variation from Non- segregating Uniform Population**: The variation occurring in non-segregating/ homogeneous population is due to environment. The environmental variation can be estimated fro the data obtained in a replicated trial from ANOVA of any standard design.
2. **Partitioning of Environmental Variation in Non- segregating Population:** The environmental variation can be separated from total by raising segregating generations *viz.*, F_2, B_1, B_2 etc along with non- segregating parents and F_1 in a same environment. The variation occurring in any segregating generation is phenotypic and the variation in any uniform population is due to environment.

$$\text{Phenotypic variance } (\sigma^2_p) = V_{F2} \text{ or } V_{F3} \text{ or } V_{B1} \text{ or } V_{B2}$$

$$\text{Environmental variance } (\sigma^2_e) = \frac{(V_{P1} + V_{P2} + 2V_{F1})}{4}$$

$$\text{Genotypic variance } (\sigma^2_g) = \sigma^2_p - \sigma^2_e$$

SOLVED EXERCISE

An experiment was conducted with 6 generations of a cross viz., P_1, P_2 F_1 F_2 B_1 and B_2. Estimate the phenotypic, genotypic and environmental variances from the following data obtained for number of pods from these generations..

Plant No.	P_1	P_2	F_1	F_2	B_1	B_2
1	50.8	40.6	48.9	52.4	49.2	46.2
2	50.3	40.2	47.8	55.4	48.3	48.0
3	53.2	42.1	49.2	49.8	47.6	45.6
4	51.6	43.3	48.2	50.2	51.2	49.2
5	50.8	40.9	47.6	53.6	54.6	50.6
6	52.3	41.2	46.8	54.2	50.2	48.2
7	50.4	39.3	48.2	51.2	49.2	47.3
8	51.6	41.5	47.3	55.0	51.0	46.0
9	52.0	42.0	49.3	49.3	48.0	47.0
10	51.0	41.0	48.6	52.3	52.5	42.2
11				58.4	48.4	48.4
12				56.2	55.0	46.0
13				60.3	54.3	50.0
14				48.2	47.2	46.2
15				55.3	53.2	47.2
16				54.2	52.3	49.3
17				50.1	52.9	45.6
18				60.3	55.5	45.5
19				63.2	53.2	43.2
20				55.2	49.3	48.2
Total	514.0	412.1	481.9	1084.8	1023.1	939.9
No. of plants	10	10	10	20	20	20

$$V_{P1} = \frac{\{\Sigma x^2 - (\Sigma x)^2}{N\}} N - 1$$

$$= \{(50.8)^2 + (50.3)^2 + ... + (51.0)^2 - (514)^2/10\}/10^{-1} = 0.84$$

Similarly,

$$V_{P2} = 1.23; \quad V_{F1} = 0.68; \quad V_{F2} = 16.30; \quad V_{B1} = 7.16;$$

and $V_{B2} = 4.47$

$$\sigma^2_e = \frac{(VP_1 + VP_2 + 2VF_1)}{4} = \frac{\{(0.84 + 1.23 + (2 \times 0.68)\}}{4}$$

$$= 0.86$$

$$\sigma^2_p = V_{F2} = 16.30 \qquad \sigma^2_g = \sigma^2_p - \sigma^2_e = 15.44$$

$$\sigma^2_p = V_{B1} = 7.16 \qquad \sigma^2_g = \sigma^2_p - \sigma^2_e = 6.30$$

$$\sigma^2_p = V_{B2} = 4.47 \qquad \sigma^2_g = \sigma^2_p - \sigma^2_e = 3.62$$

Conclusion: From the estimates of variances of different segregating generations revealed that phenotypic and genotypic variances for F_2 was the highest followed by variances of B_1 and B_2, indicating that the variability in F_2 generation was higher than B_1 and B_2

3. **Partitioning of $g \times e$ interaction component:** The pooled analysis of variance from the replicated data of trials conducted at different locations/seasons is a prerequisite for estimation of the interaction component of variation in a quantitative character.

SOLVED EXERCISE

The following is a data recorded on number of bolls of 8 cotton genotypes replicated thrice in three locations. Partition the variance into different components.

Geno-type	Location–1				Location-2				Location-3			
	R-I	R-II	R-III	Total	R-I	R-II	R-III	Total	R-I	R-II	R-III	Total
1	32.7	35.2	38.2	106.1	52.3	50.3	54.2	156.8	R-I	R-II	R-III	Total
2	26.6	28.4	26.3	81.3	48.3	46.2	45.3	139.8	42.2	40.5	44.2	126.9
3	27.8	29.7	28.3	85.8	46.1	44.4	48.3	138.8	38.2	36.3	35.3	109.8
4	21.7	22.3	24.3	68.3	36.3	34.3	33.1	103.7	36.6	34.8	38.3	109.7
5	40.0	38.2	37.4	115.6	55.2	51.6	53.2	160.0	34.3	30.3	32.2	96.8
6	38.0	35.2	36.3	109.5	48.3	46.2	45.4	139.9	45.2	41.6	43.6	130.4
7	36.2	35.8	37.3	109.3	45.6	47.3	46.3	139.2	35.9	30.6	33.6	100.1
8	35.3	33.2	32.1	100.6	44.9	42.3	45.3	132.5	38.6	37.3	36.9	112.8
Total	258.3	258.0	260.2	776.5	377.0	362.6	371.1	1110.7	40.3	40.1	41.6	122.0

ANOVA for individual location

Sources	*df*	MS			
		Location–1	Location-2	Location-3	
Replications	2	0.18	6.55	13.02	MS_1
Genotypes	7	92.61	97.50	48.78	MS_2
Error	14	2.65	2.13	1.63	MS_3

Individual Location

Location 1:

$$\text{Environmental variance} = \sigma^2_e = 2.65$$

$$\text{Genotypic variance} = \sigma^2_g = \frac{(MS_2 - MS_3)}{r} = \frac{(92.61\text{-}2.65)}{3} = 29.99$$

$$\text{Phenotypic variance} = \sigma^2_p = \sigma^2_g + \sigma^2_e = 29.99 + 2.65 = 32.64$$

Location 2:

$$\text{Environmental variance} = \sigma^2_e = 2.13$$

$$\text{Genotypic variance} = \sigma^2_g = \frac{(MS_2 - MS_3)}{r} = \frac{(97.50 - 2.13)}{3} = 31.79$$

$$\text{Phenotypic variance} = \sigma^2_p = \sigma^2_g + \sigma^2_e = 31.79 + 2.13 = 33.92$$

Location 3:

$$\text{Environmental variance} = \sigma^2_e = 1.63$$

$$\text{Genotypic variance} = \sigma^2_g = \frac{(MS_2 - MS_3)}{r} = \frac{(48.78\text{–}1.63)}{3} = 15.72$$

$$\text{Phenotypic variance} = \sigma^2_p = \sigma^2_g + \sigma^2_e = 15.72 + 1.63 = 17.35$$

Two way table

Genotype	**Location**			
	1	**2**	**3**	**Total**
1	106.1	156.8	126.9	389.8
2	81.3	139.8	109.8	330.9
3	85.8	138.8	109.7	334.3
4	68.3	103.7	96.8	268.8
5	115.6	160.0	130.4	406
6	109.5	139.9	100.1	349.5
7	109.3	139.2	112.8	361.3
8	100.6	132.5	122.0	355.1
Total	776.5	1110.7	908.5	2795.7

$$\text{Correction Factor} = \frac{(2795.7)^2}{24} = 325664.1$$

$$\text{Total SS} = (106.1^2 + ... + 122.0^2) - CF = 12100.09$$

$$\text{Genotype SS} = \frac{(389.8^2 + ... + 355.1^2)}{3} - CF = 4025.473$$

$$\text{Location SS} = \frac{(776.5 + ... + 908.5^2)}{8} - CF = 7083.270$$

Genotype × Location SS = Total SS – Genotype Ss – Location SS = 991.34

Pooled ANOVA Table

Sources	***df***	**SS**	**MS**	**F ratio**
Genotypes	7	4025.47	575.06 (MS_1)	MS_1/MS_4= 269.9*
Locations	2	7083.27	3541.63 (MS_2)	MS_2/MS_4= 1654.9*

Contd...

Contd...

Genotype × Location	14	991.34	70.81 (MS_3)	MS_3/MS_4= 33.1*
Pooled error (14 x3)	42	89.74	2.14 (MS_4)	

Pooled error SS = MS (L_1) × error *df* + MS (L_2) × error *df* + MS (L_3) × error df

= (2.65 × 14) + (2.13 × 14) + (1.63 × 14) = 89.74

Pooled error variance = σ^2_e = 2.14

$$\text{Genotypic variance} = \sigma^2_g = \frac{(MS_1 - MS_3)}{r_1} = \frac{(575.06\text{-}70.81)}{3} \times 3 = 56.03$$

$$\text{Location variance} = \sigma^2_1 = \frac{(MS_1 - MS_3)}{rg} = \frac{(3541.63\text{-}70.81)}{3} \times 8 = 144.62$$

$$\text{Genotype} \times \text{Location interaction} = \sigma^2_{gl} = \frac{(MS_3 - MS_4)}{r} = \frac{(70.81 - 2.14)}{3} = 22.89$$

CONCLUSION

From the pooled ANOVA, it is revealed that MSS due to genotypes and locations were significant indicating there was variability among genotypes and locations. Further, the significance of G × L indicated that genotypes responded differentially in different locations.

8 Correlation Coefficients

Correlation refers to the degree and direction of relationship between two or more variables. Correlation coefficient is a statistical measure which is used to find the degree and direction of association between two characters in plant breeding experiments. The intensity of correlation coefficient is represented as 'r' and ranges from –1 to +1. It is independent of unit of measurement. If $r = -1$, it indicates the relation between two characters in opposite direction, *i.e.*, the high value of one character is associated with low value of other character. If $r = +1$, then complete perfect association between the two characters varying in the same direction is predicted. If $r = 0$, there is no correlation between the variables and the variances are independent of each other. In plant breeding simple correlation coefficient is measured to find out mutual relationship between two plant characters and to determine the component characters, on which selection can be done for genetic improvement in yield. It is of three types, *i.e.*, phenotypic, genotypic and environmental.

The estimation of genotypic, phenotypic and environment population coefficient consists of following steps

1. Evaluation of genotypes in replicated trails
2. Performing ANOVA and ANCOVA for different characters and their gene pairs
3. Estimation of variances and co variances
4. Estimation of different correlation coefficients

Phenotypic correlation: Refers to the observable correlation between two characters and is characterized as

$$\text{Phenotypic correlation coefficients, } rPxy = \frac{CoVP(x,y)}{\sqrt{VP(x) \cdot VP(y)}}$$

where, $CoVP(x, y)$ = Phenotypic covariance between two traits X and Y

VP(x) = Phenotypic variance for X and = Phenotypic variance for Y

Genotypic correlation: Refers to the heritable association between two characters and is characterized as

$$\text{Genotypic correlation coefficients, } rG\,xy = \frac{CoVg(x,y)}{\sqrt{VG(x) \cdot VG(y)}}$$

where, $CoVG(x, y)$ = genotypic covariance between two traits X and Y

VG(x) = genotypic variance for X and = genotypic variance for Y

Environmental correlation: It is the association of two traits due to environmental effects and is the ratio of environmental covariance, two variables x & y to the environmental standard deviation and calculated as

$$\text{Environmental correlation coefficients, } rExy = \frac{CoVE(x,y)}{\sqrt{VE(x) \cdot VE(y)}}$$

where, $CoVE(x, y)$ = Environmental covariance between two traits X and Y

VE(x) = Environmental variance for X and

VE(y) = Environmental variance for Y

The significance of phenotypic and genotypic correlation coefficients was tested against 'r' values given by Snedcor and Cochran (1967) at (n – 2) degrees of freedom at 5 per cent and 1 per cent levels of significance.

$$t \text{ cal} = \frac{r\sqrt{n-2}}{\sqrt{1-r^2}}$$

Where,

r = Correlations coefficient

n = Total no. of pairs of observations

9 Path Coefficient Analysis

To establish a cause and effect relationship, the first step used was to partition the genotypic and phenotypic correlation coefficients into direct and indirect effects by path analysis as suggested by Dewey and Lu (1959).

The second step in the path analysis is to prepare path diagram based on cause and effect relationship. In the present study, path diagram was prepared by taking yield as the effect *i.e.* functions of various components like x_1, x_2, x_3........... and these components showed following type of association with each other.

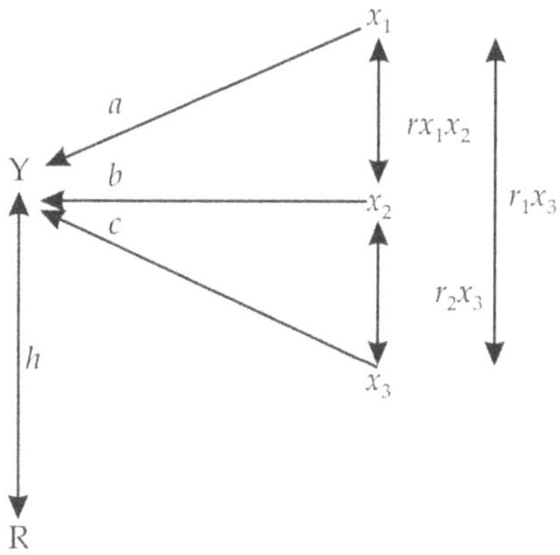

Path diagram

In the path diagram the yield is the result of x_1, x_2, x_3 and some other undefined factors designated by R. The double arrow lines indicate mutual associations as measured by correlation coefficients. The single arrow represents direct effect as measured by path coefficient Pij. In the Figure *a*, *b*, *c* and *h* are the path coefficients due to respective variables.

Path coefficient were obtained by solving a set of simultaneous equations as given below,

$$r(x_1,y) = a + r(x_1 x_2)\, b + r(x_1x_3)\, c$$
$$r(x_2,y) = r(x_2x_1)\, a + b + r(x_2x_3)\, c$$
$$r(x_3,y) = r(x_3x_1)\, a + r(x_3x_2)\, b + c$$
$$r(R,y) = h$$

Where,

a = Direct effect of x_1 on yield

b = Direct effect of x_2 on yield
c = Direct effect of x_3 on yield
h = Direct effect of undefined factor (R)
$r(x_1 y)$ = Correlation between x_1 and yield
$r(x_2 y)$ = Correlation between x_2 and yield
$r(x_3 y)$ = Correlation between x_3 and yield
$r(Ry)$ = Correlation between R and yield
$r(x_1x_2)b$ = Indirect effect on yield via x_2 (For character x_1)
$r(x_1x_3)c$ = Indirect effect on yield via x_3 (For character x_1)
$r(x_1x_2)a$ = Indirect effect on yield via x_1 (For character x_2)

Considering only the first three factors *i.e.* X_1, X_2 and X_3 the simultaneous equations given above can be presented in matrix notations as :

$$\underset{A}{\begin{pmatrix} rx_1y \\ rx_2y \\ rx_3y \end{pmatrix}} = \underset{B}{\begin{pmatrix} rx_1x_1 & rx_1x_2 & rx_1x_3 \\ rx_1x_2 & rx_2x_2 & rx_2x_3 \\ rx_1x_3 & rx_3x_2 & rx_3x_3 \end{pmatrix}} \times \underset{C}{\begin{pmatrix} a \\ b \\ c \end{pmatrix}}$$

$$A = B.C \qquad C = B^{-1}A$$

After having calculated the values of path coefficients, *i.e.* 'C' matrix, it is possible to obtain the path value for residual (R) in the following way.

$$1 = R^2 + a^2 + b^2 + c^2 + 2r(x_1x_2)ab + 2r(x_1x_3)ac + 2r + (x_2x_3)bc$$

$$R^2 = 1 - [a^2 + b^2 + c^2 + 2r(x_1x_2)ab + 2r(x_1x_3)ac + 2r + (x_2x_3)bc]$$

$$R = \sqrt{1 - a^2 - b^2 - c^2 - 2r(x_1x_2)ab - 2r\,(x_1x_3)ac - 2r(x_2x_3)bc}$$

SOLVED EXERCISE

In an experiment 10 genotypes of Pearl millet were evaluated in RBD with 3 repl. The mean values over 3 replications for No. of tillers, earhead girth, earhead length and grain yield/plant (g) are as below. Compute Correlation and Path Coefficient and interpret the results.

From the above data, analysis of variance for four characters and analysis of covariance for six combination of characters were carried out and the mean squares and mean products are given below.

Mean squares for four characters in Pearl millet

	Mean squares (Variance)			
	X_1	X_2	X_3	Y
Phenotypic	0.123	1.477	2.435	30.121
Genotypic	0.11	1.171	1.226	27.832
Error	0.014	0.307	1.209	2.289

x_1: No. of tillers/plant

Geno.	RI	RII	RIII	Total	Mean
1	3.20	3.36	3.26	9.82	3.27
2	3.02	3.20	3.13	9.35	3.12
3	3.82	3.90	4.12	11.84	3.95
4	3.52	3.63	3.76	10.91	3.64
5	3.62	3.54	3.71	10.87	3.62
6	3.92	4.08	3.84	11.84	3.95
7	3.24	2.86	3.08	9.18	3.06
8	3.62	3.58	3.54	10.74	3.58
9	3.58	3.68	3.77	11.03	3.68
10	3.94	3.88	4.15	11.97	3.99

x_2: Ear head girth (cm)

Geno.	RI	RII	RIII	Total	Mean
1	10.20	12.10	11.40	33.70	11.23
2	9.50	10.30	10.00	29.80	9.93
3	12.50	13.20	14.50	40.20	13.40
4	12.10	12.80	13.20	38.10	12.70
5	11.98	11.30	12.64	35.92	11.97
6	13.60	14.00	13.20	40.80	13.60
7	10.60	10.40	11.50	32.50	10.83
8	11.60	12.30	11.90	35.80	11.93
9	11.50	12.60	13.20	37.30	12.43
10	11.64	11.20	12.60	35.44	11.81

x_3: Ear head length (cm)

Geno	RI	RII	RIII	Total	Mean
1	20.40	24.10	22.40	66.90	22.30
2	22.10	22.90	23.00	68.00	22.67
3	21.92	22.60	23.86	68.38	22.79
4	23.40	25.40	26.80	75.60	25.20
5	23.80	23.40	25.90	73.10	24.37
6	23.90	25.80	24.60	74.30	24.77
7	21.40	20.90	22.50	64.80	21.60
8	22.60	22.20	20.90	65.70	21.90
9	22.60	23.90	24.60	71.10	23.70
10	24.30	22.60	26.20	73.10	24.37

Y : Grain yield /plant(g)

Geno	RI	RII	RIII	Total	Mean
1	62.10	66.40	65.30	193.80	64.60
2	62.30	65.80	67.10	195.20	65.07
3	68.20	69.90	71.30	209.40	69.80
4	72.10	75.20	76.60	223.90	74.63
5	72.60	75.80	77.40	225.80	75.27
6	72.60	74.10	71.90	218.60	72.87
7	63.30	60.90	65.40	189.60	63.20
8	69.90	72.10	70.80	212.80	70.93
9	76.50	78.90	79.40	234.80	78.27
10	76.30	74.20	79.30	229.80	76.60

Mean products for six combinations of four characters in Pearl millet

	Mean product (Covariance)					
	X_1X_2	X_1X_3	X_1Y	X_2X_3	X_2X_Y	X_3Y
Phenotypic	0.356	0.337	1.465	1.108	3.942	6.06
Genotypic	0.306	0.242	1.334	0.625	3.384	4.822
Error	0.050	0.096	0.131	0.483	0.558	1.238

Simple correlation coefficients are calculated with the following formula.

$$rx_1x_2 = \frac{Cov(x1x2)}{\sqrt{V(x1)\cdot V(x2)}}$$

$$\textbf{Phenotypic} = \frac{CovP(x1x2)}{\sqrt{VP(x1)\cdot VP(x_2)}} = \frac{0.356}{\sqrt{0.123\times 1.477}} = 0.84$$

$$\textbf{Genotypic} = \frac{CovG(x1x2)}{\sqrt{VG(x1)\cdot VG(x2)}} = \frac{0.306}{\sqrt{0.110\times 1.171}} = 0.85$$

$$\textbf{Environment} = \frac{CovE(x1x2)}{\sqrt{VE(x1)\cdot VE(x2)}} = \frac{0.050}{\sqrt{0.014\times 0.307}} = 0.77$$

Similarly correlation coefficients at phenotypic and genotypic levels are computed and given below in matrix form.

Phenotypic correlation

	x_1	x_2	x_3	y
x_1	**1.00**	0.84	0.62	0.76
x_2	0.84	**1.00**	0.58	0.59
x_3	0.62	0.58	**1.00**	0.71
y	0.76	0.59	0.71	**1.00**

Genotypic Correlation

	x_1	x_2	x_3	y
x_1	**1.00**	0.85	0.66	0.76
x_2	0.85	**1.00**	0.52	0.59
x_3	0.66	0.52	**1.00**	0.83
y	0.76	0.59	0.83	**1.00**

PATH COEFFICIENTS (GENOTYPIC)

Simultaneous equations for computing genotypic paths are

$$r\,x_1y = r\,x_1\,x_1a + r\,x_1\,x_2b + r\,x_1\,x_3c$$
$$r\,x_2y = r\,x_2\,x_1a + r\,x_2\,x_2b + r\,x_2\,x_3c$$
$$r\,x_3y = r\,x_3\,x_1a + r\,x_3\,x_2b + r\,x_3\,x_3c$$

$$\underset{A}{\begin{pmatrix} rx_1y \\ rx_2y \\ rx_3y \end{pmatrix}} = \underset{B}{\begin{pmatrix} rx_1x_1 & rx_1x_2 & rx_1x_3 \\ rx_1x_2 & rx_2x_2 & rx_2x_3 \\ rx_1x_3 & rx_3x_2 & rx_3x_3 \end{pmatrix}} \times \underset{C}{\begin{pmatrix} a \\ b \\ c \end{pmatrix}}$$

$$A = B \cdot C \quad i.e., \quad C = B^{-1}A$$

Obtain inverse of matrix B

The matrix with genotypic correlation is taken as B matrix.

$$B^{-1} = \begin{pmatrix} 4.71 & -3.27 & -1.41 \\ -3.27 & 3.64 & 0.26 \\ -1.41 & 0.26 & 1.79 \end{pmatrix}$$

$$C = B^{-1}A \quad \begin{pmatrix} 4.71 & -3.27 & -1.41 \\ -3.27 & 3.64 & 0.26 \\ -1.41 & 0.26 & 1.79 \end{pmatrix} \times \begin{pmatrix} 0.76 \\ 0.59 \\ 0.83 \end{pmatrix}$$

$$a = (4.71 \times 0.76) + (-3.27 \times 0.59) + (-1.41 \times 0.83) = 0.481$$

$$b = (-3.27 \times 0.76) + (3.64 \times 0.59) + (0.26 \times 0.83) = -0.117$$

$$c = (-1.41 \times 0.76) + (0.26 \times 0.59) + (1.79 \times 0.83) = 0.573$$

Residual effect

$$1 = R^2 + (0.481)^2 + (-0.117)^2 + (0.573)^2 + 2\,(0.85)\,(0.481)\,(-0.117) + 2\,(0.66)\,(-0.117)\,(0.83) + 2\,(0.52)\,(-0.117)\,(0.573)$$

$$= R^2 + 0.772$$

$$R = \sqrt{1-0.772} = 0.48$$

Direct and indirect effects:

1. x_1 and y — Direct (a) = a = 0.481
   Indirect via $x_2 = brx_1x_2 = (-0.117)\,(0.85) = -0.100$
   Indirect via $x_3 = crx_1x_3 = (0.573)\,(0.66) = 0.378$
2. x_2 and y — Direct (b) = b = – 0.117
   Indirect via $x_1 = arx_1x_2 = (0.481)\,(0.85) = 0.409$
   Indirect via $x_3 = crx_2x_3 = (0.573)\,(0.52) = 0.298$
3. x_3 and y — Direct (c) = c = 0.573
   Indirect via $x_1 = arx_1x_3 = (0.481)\,(0.0.66) = 0.317$
   Indirect via $x_2 = brx_2x_3 = (-0.117)\,(0.0.52) = -0.061$

Phenotypic (above diagonal) and Genotypic (below diagonal) correlation coefficients

Characters	No. of tillers (x_1)	Earhead girth (x_2)	Earhead length (x_3)	Grain yield (y)
No. of tillers (x_1)	–	0.84**	0.62**	0.76**
Earhead girth (x_2)	0.85**	–	0.58**	0.59**
Earhead length (x_3)	0.66**	0.52**	–	0.71**
Grain yield (y)	0.76**	0.59**	0.83**	–

Genotypic path coefficients of three characters and correlation with yield

Characters	No. of tillers/ plant (x_1)	Earhead girth (x_2)	Earhead length (x_3)	Correlation with grain yield (y)
No. of tillers (x_1)	**0.481**	–0.100	0.378	0.760**
Earhead girth (x_2)	0.409	**–0.117**	0.298	0.590**
Earhead length (x_3)	0.318	–0.061	**0.573**	0.830**

Residual effect (R) = 0.48

CONCLUSION

From the correlation coefficient it is revealed that no. of tillers, earhead girth and earhead length had highly significant correlation with grain yield at both phenotypic and genotypic levels. Among the yield contributing characters, all characters had highly positive significant correlation among themselves and hence these characters were inter related.

From the path analysis, it is revealed that earhead length and no. of tillers had the highest positive direct effect on grain yield whereas earhead girth had negative direct effects. However, correlation of ear head with with grain yield is positive and significant. This may be due to indirect effect via no. of tillers and earhead length. Hence, selection of no of tillers and earhead length may be directly made for increased grain yield in pearl millet.

Residual effect was high since the characters contributed only 52.0 % to the total variability in grain yield. It indicated that other characters which have not been studied here, needs to be included in this analysis to account fully for the variation in grain yield.

10

Gene Action

NATURE OF GENE ACTION

The first attempt to partition the genotypic variation in polygenic traits into its component part was made by Fisher (1918). There are three components of hereditary variation.

1. **Additive variance:** It is that portion of genetic variance which results due to average effect of genes on all segregating loci. Thus it is the component which arises from the difference between homozygotes of a gene, AA or aa. It is a measure of additive gene action and additive genes show intermediate expression of two homozygotes. Additive variance is fixable and responds to selection, therefore, selection for traits governed by such variance is very effective. Additive gene action is the chief cause of resembles between relatives and progress by selection is directly proportional to the degree of resemblance between parent and its progeny. Thus, additive gene action is a measure of breeding value of genotype. Transgressive ssegregant is the result of additive genetic variance and the combining ability effects of parent is a measure of additive gene effects.
2. **Dominance variance:** A dominance the component arising from interaction of alleles (Intra–allelic interaction). It is due to deviation of heterozygote (Aa) from the average of two homozygotes (AA and aa). It is a measure of dominance gene action. Dominance variance is associated with heterozygosity and, therefore, it is expected to be maximum in cross pollinated crops and minimum in self pollinated crop species. It is not flexible and hence selection for traits controlled by such variances is not effective. Dominance variance is the chief cause of hetrerosis and specific combining ability effect of cross is a measure of dominant variance.
3. **Epistatic variance:** An epistatic part associated with interaction of non alleles (inter–allelic interaction).Epistatic variance includes both additive and non additive components. It is of three types *viz.*, Additive × additive, additive × dominance and dominance × dominance.
   - (*i*) *Additive × Additive:* It refers to interaction between two or more loci each exhibiting lack of dominance individually. It is fixable and selection is effective for traits governed by such variance.
   - (*ii*) *Additive × Dominance:* It refers to interaction between two or more loci, one exhibiting lack of dominance and other dominance individually.

(*iii*) *Dominance* × *Dominance:* It refers to interaction between two or more loci each exhibiting dominance individually.

Both additive × Dominance and Dominance and Dominance variances are non–fixable and hence heterosis breeding and recurrent selection for SCA may be rewarding for traits exhibiting such variances.

There are two basic phenomena concerned with metric characters.

1. **The resemblance between relatives:** Relatives tend to be resemble each other, and the closer the relationship, in general the closer the resemblance. The degree of relationship between relatives is one of the properties of a population that can be readily observed. The resemblance between off-spring and parents provides the basis for selective breeding.
2. **The inbreeding depression:** Inbreeding tends to reduce the mean level of all characters closely connected with fitness in naturally out breeding plants, and in consequences to loss of general vigour and fertility.

The properties of population that can observe in connection with metric charc-ters are mean, variances and covariances. The natural subdivision of the population into families allows us to anlalyse the variance into components which forms the basis of the measurement of the degree of rsemblance between relatives

3. **Values and Means:** The value observed when the character is measured on an individual is the 'phenotypic' value of that individual. All observations *i.e.* mean, variances or covariances are based on phenotypic values. Genetic properties of the population are then worked out by dividing the phenotypic value into its components.

$$P = G + E$$

where, P = Phenotypic value; G = Genotype; E = Environment

Thus the two components of value associated with genotype and environment are the genotypic value and environment deviation.

Consider a single locus with two alleles A_1 and A_2 , a population will having three genotypes A_1A_1, A_1A_2 and A_2A_2. The genotypic value of one homozygote '$+a$', that of other homozygote '$-a$' and that of heterozygote 'd'. Then can draw a scale diagram as

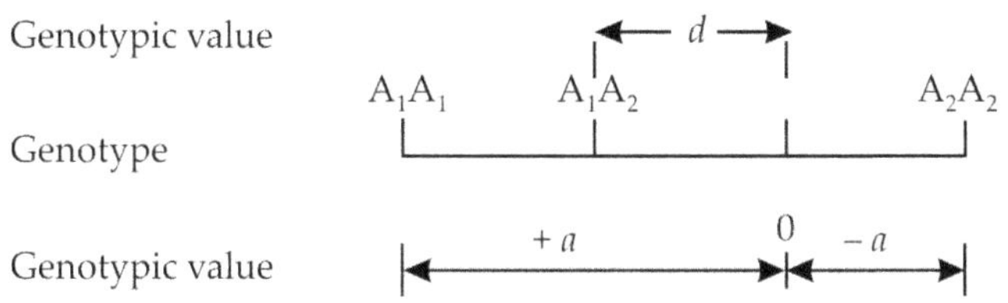

Fig: 10.1 Arbitrarily assigned genotypic value

The origin on this scale is midway between the values of two homozygotes. The value 'd' of heterozygote depends on the degree of dominance. *i.e*

Absence of dominance	$d = 0$
A_1 is of dominant	$d = +ve$

A_2 is of dominant	$d = -ve$
Complete dominance of A_1 or A_2	$d = +a$ or $-a$
Over dominance of A_1 or A_2	d= > +a or d= < –a
The degree of dominance	= d/a

POPULATION MEAN

Consider a single locus situation with two alleles viz., A_1 and A_2. With respect to this locus there will be three genotypes viz., A_1A_1, A_1A_2 and A_2A_2 in the population. The frequency of these genotypes in a large random mating population will be P^2, $2pq$ and q^2, respectively. Population mean is therefore, sum of the product of frequency of the genotype and its value, summations being made over all the genotypes.

Genotype	Frequency	Value
A_1A_1	P^2	a
A_1A_2	$2pq$	d
A_2A_2	q^2	$-a$

$$\text{Population mean} = P^2a + 2pqd - q^2a$$
$$= a\,(P^2 - q^2) + 2pqd$$
$$= a\,[(p + q)\,(p - q)] + 2pqd$$

Since $p + q = 1,$

$$\text{Population mean} = a\,(p - q) + \mathbf{2}pqd$$

This is genotypic as well as phenotypic mean

If there is no dominance *i.e.* $d = 0$, the mean is proportional to gene frequency, then population mean $= a\,(p - q)$

$$= a\,(1 - q - q) \quad (\text{since } P = 1 - q)$$
$$= a\,(1 - 2q)$$

If A_1 is completely dominant then population mean

$$= a\,(p - q) + 2apq \qquad (\text{as } d = a)$$
$$= a\,(1 - 2q) + 2\,aq\,(1 - q) \qquad (\text{since } p = 1 - q)$$
$$= a\,(1 - 2q) + 2aq - 2aq^2$$
$$= a\,(1 - 2q + 2q - 2\,q^2)$$
$$= a\,(1 - 2\,q^2)$$

Thus, the mean in this case is proportional to square of gene frequency

How population mean is affected by the gene frequency?

Assume that the genotypic values of genotypes A_1A_1, A_1A_2 and A_2A_2 are 16, 12 and 4, respectively. Suppose the frequency of gene A_2 is 0.4

Frequency of A_1 = 1– frequency of A_2 = 1.0 – 0.4 = 0.6

Genotype	Frequency	Value
A_1A_1	0.36	16
A_1A_2	0.48	12
A_2A_2	0.16	4

The mid point $= \frac{16+4}{2} = 10$

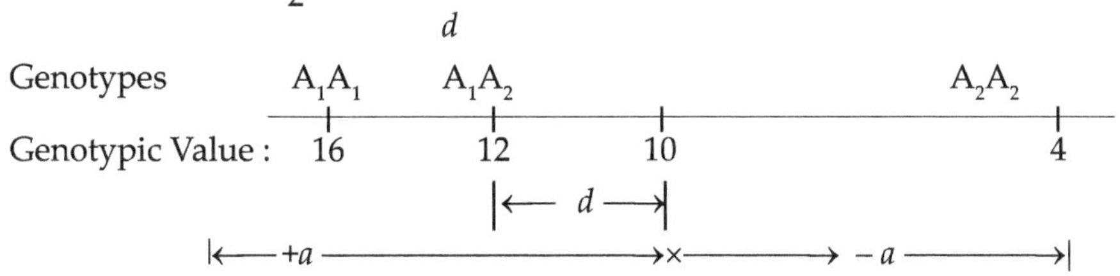

$$a = 16 - 10 = 6$$
$$d = 12 - 10 = 2$$

Now, Population mean $= a\,(p - q) + 2pqd$

$$= 6\,(0.6 - 0.4) + 2\,(0.6)\,(0.4) \times 2$$
$$= 6\,(0.2) + 0.96 = 2.16$$

Since the values of '*a*' and '*d*' are obtained as deviation from the mid–point (*i.e.* 10) then

Population mean = $10 + 2.16 = 12.16$

Now suppose the frequency of A_2 = 0.9

Then the frequency of A_1 = 0.1

Population mean = $6\,(0.1 – 0.90) + 2 \times 0.10 \times 0.90 \times 2$

$= 6\,(–0.80) + 4\,(0.09)$

$= – 4.80 + 0.36 = –4.44$

Hence, population mean = 10 – 4.44 = 5.56

The population mean has now declined as the frequency of allele with negative effect on the character has risen.

AVERAGE EFFECT

As parents passed on their genes and not their genotypes to the next generation, genotypes being created afresh in each generation, a new measure of value that refers to the genes is needed. The new measures of value associated with genes are known as the average effect. Average effect depends on the genotypic values ('*a*' and '*d*'), and on the gene frequencies. Average effect is therefore, properties of population as well as genes concerned. Average effect enables to assign a 'breeding value' to the individuals.

Definition: Average effect of a gene (allele) is the mean deviation from population mean of individuals which received that gene from one parent, the gene received from other parent having come at random from the population.

Consider a single locus with two alleles *i.e.* A_1 and A_2. If an individual A_1A_1 is matted randomly to a number of individuals in a population, then the gametes produced by A_1A_1 all carrying A_1 gene will unite at random with the gametes from the population. Then the mean of the genotypes so produced deviates from the population mean by an amount which is the average effect of A_1 gene. The average effect of A_1 is denoted by α_1. The three possible genotypes with respect to this locus in the population produces two types of gametes *i.e.* A_1 and A_2 with frequencies p and q respectively. If the gametes (all A_1) produces by A_1A_1 individual unite randomly with the gametes from the population, two kinds of individuals will be produced *i.e.* A_1A_1 and A_1A_2 with the frequencies p and q respectively.

Suppose the genotypic value of A_1A_1 is 'a' and genotypic value of A_1A_2 is 'd'. The mean genotypic value of the progeny will therefore be $pa + qd$.

Deviation of this mean genotypic value from the population mean is the average effect of the gene genotypic value of A_1 *i.e.* α_1.

$$\begin{aligned}\alpha_1 &= (pa + qd) - [a(p - q) + 2pqd]\\ &= (pa + qd) - [pa - qa + 2pqd]\\ &= pa + qd - pa + qa - 2pqd\\ &= q\,[a + d - 2pd]\\ &= q\,[a + d\,(1 - 2p)]\\ &= q\,[a + d\,(1 - p - p] \qquad \text{since } 1 - p = q\\ \boldsymbol{\alpha_1} &= \boldsymbol{q\,[a + d\,(q - p]}\end{aligned}$$

Similarly, average effect of A_2 *i.e.* α_2

$$\boldsymbol{\alpha_2 = -p\,[a + d\,(q - p]}$$

Difference between α_1 and α_2 is $[a + d(q - p]$ which is the average effect of gene substitution (α)

Since α is $[a + d\,(q - p)]$ we can write

$$\alpha_1 = q\alpha$$
$$\alpha_2 = -p\alpha$$

BREEDING VALUE

The value of an individual judged by the mean value of its progeny is called the breeding value of the individual. Breeding value, unlike average effect can, therefore, be measured. If an individual is mated to number of individuals taken at random from the population, then its breeding value is twice the mean deviation of the progeny from the population mean. The deviation has to be doubled be cause the parent in question provides only half of the genes in the progeny, the others half coming at random from the population. Breeding value is usually expressed in the form of deviation from the population mean. Just as the average effect is a property of the gene and the population, so is the breeding value a property of the individual and the population from which its mates are drawn.

In terms of average effects, the breeding value of an individual is equal to the sum of average effects of the genes it carries, the summation being made over the pair of alleles at each locus and over all the loci assuming of non–allelic interactions.

Assuming a single locus with two alleles the breeding value of the genotypes will be

Genotype		Breeding value	Frequency
A_1A_1	$(2\alpha_1)$	$2q\alpha$	p^2
A_1A_2	$(\alpha_1+\alpha_2))$	$(q-p)\,\alpha$	$2pq$
A_2A_2	$(2\alpha_2)$	$-2p\alpha$	q^2

When several loci are considered, breeding value of an individual is the sum of the breeding values over all loci, in the absence of non– allelic interactions.

As the average effects are expressed as deviations from population mean, average breeding value in the population is zero.

$$\text{Mean breeding value in the population} = p^2(2q\alpha)+2pq((q\alpha-p\alpha)-2pq2\alpha$$
$$= 2p^2q\alpha+2pq^2\alpha-2p^2q\alpha-2pq2\alpha$$
$$= 0$$

Breeding value is referred to as additive genotypes and the variance of breeding value as additive variance.

DOMINANCE DEVIATION

Non additivity of two alleles at locus is the cause of dominance deviation.

When a single locus is under consideration, the difference between the genotypic value 'G' and the breeding value 'A' of a particular genotype is known as dominance deviation.

So, $G = A + D$ $D = G - A$

From statistical point of view, the dominance deviations are interactions between allels or within locus interactions. They represent the effect of putting two alleles in pairs to make genotype; the effect not accounted for by the effect of two genes taken singly. The dominance deviations are influenced by the frequencies well as 'd'. If there is no dominance *i.e.* d =0, the dominance deviations are also zero, and then breeding values and genotypic values are the same. Genes that show no dominance are called additive genes or are said to act additivity.

EXAMPLE

The dwarfing gene pygmy (*pg*) is a recessive gene that reduces the weight of body. Given the following, calculate population mean, average effects, breeding values and dominance deviation if $q = 0.1$ and $q = 0.4$

	Genotype		
	++	*+pg*	*pgpg*
Body weight	14	12	6

Solution:

$$\text{Mid point} = \frac{14+6}{2} = 10$$

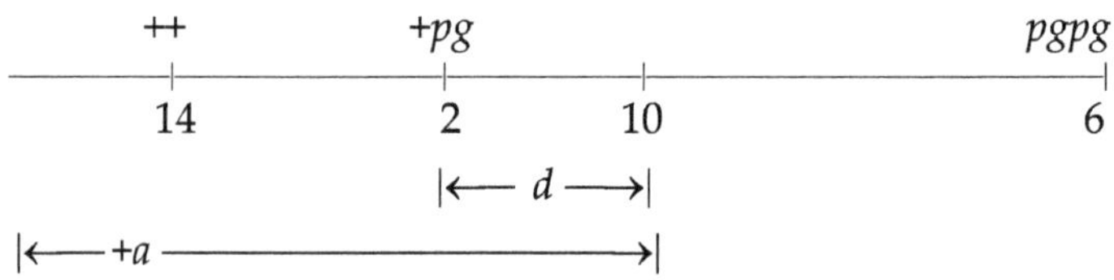

Thus,

$$a = 14 - 10 = 4\text{g}$$
$$d = 12 - 10 = 2\text{g}$$

Population mean:

When $q = 0.1 \;\; ; p = 1 - 0.1 = 0.9$

$$\mu = a\,(p - q) + 2pqd$$
$$= 4\,(0.8) + 2\,(0.18) \quad = 3.56$$

$$\text{Population mean} = 10 + 3.56 = 13.56$$

When $q = 0.4 \;\; ; p = 1 - 0.4 = 0.6$

$$\mu = a\,(p - q) + 2pqd$$
$$= 4\,(0.2) + 2\,(0.48) = 1.76$$

$$\text{Population mean} = 10 + 1.76 = 11.76$$

Average effect of gene substitution (α)

$$\alpha = q + d\,(q - p]$$

When $q = 0.1$ When $q = 0.4$

$\alpha = 4 + 2\,(0.1 - 0.9)$ $\alpha = 4 + 2\,(0.4 - 0.6)$

$\alpha = 2.4$ $\alpha = 3.6$

Now average effect of + *i.e* α_1 and *pg i.e.* α_2 will be estimated using

$$\alpha_1 = q\alpha$$
$$\alpha_2 = -p\alpha$$
$$\alpha = \alpha_1 - \alpha_2$$

	$q = 0.1$	$q = 0.4$
α_1	0.24	1.44
α_2	–2.16	–2.16
α	2.40	3.60

Breeding value

	Mean	Breeding value		
		++	+*pg*	*pgpg*
Formula		$2q\alpha$	$(q-p)\,\alpha$	$-2q\alpha$
$q = 0.1$	13.56	0.48	–1.92	–4.32
$q = 0.4$	11.76	2.88	–0.72	–4.32

Genotypic values and Dominance Deviations

First calculate genotypic values (G) and then calculate Dominance deviations (D) as

D = G–A

	$q = 0.1$ & $\mu = 13.56$			$q = 0.4$ & $\mu = 13.56$		
	++	+*pg*	*pgpg*	++	+*pg*	*pgpg*
Frequency	0.81	0.18	0.01	0.36	0.48	0.16
Genotypic value (G)	0.44	–1.56	–7.58	2.24	0.24	–5.76
Breeding value (A)	0.48	–1.92	–4.32	2.88	–0.72	–4.32
Dominance Deviation (D)	–0.04	0.36	–3.24	–0.64	0.96	–1.44

INTERACTION DEVIATION

At a single locus, the genotypic value is made up of the breeding value and dominance deviation only. But when the genotype refers to more than one locus, the genotypic value may contain an additional deviation due to non additive combination.

Suppose

G_A = the genotypic value of an individual attributable to one locus

G_B = the genotypic value of an individual attributable to second locus

G = aggregate genotypic value attributable to both loci together

Then, $G = G_A + G_B + I_{AB}$

Where, I_{AB} is the deviation from additive additive combination of these genotypic values.

If I is not zero for any combination of genes at different loci, those genes are said to interact or exhibit epistasis. The deviation I is called the interaction deviation or epistatic deviation. Loci may interact in pairs or in three or higher numbers and interaction may be of several different sorts, as the behavior of major genes shows. All loci together

$$G = D + H + I$$

Where, D = sum of breeding value attributable to the separate loci

H = sum of dominance deviations

The mean interaction deviation of all the genotypes in a population is zero, when values are expressed as deviations from the population mean. The interaction deviation is depends on the frequency of the genotypes in population and so on the gene frequencies.

The values and means may be assigned to different experimental population. Several authors (Fisher et al., 1932., Mather, 1949 and Mather and Jinks, 1971) used genetic models to describe gene action of any number of genes on a given character. Let us assume that there are two extreme parents P_1 and P_2 with genotypes with AA and aa , respectively with respect to gene A. then the mean of parents and their F_1 will have the genotypic value as follows.

Without non–allelic interaction

$$\bar{P}_1(AA) = m + d$$
$$\bar{P}_2\ (aa) = m - d$$
$$\bar{F}_1\ (Aa) = m \pm h$$

Where,

m is the mid–point

d is the genotypic value of the homozygote

h is the genotypic value of the heterozygote

For F_2 population, the phenotypic value will include

$$\bar{F}_2 = \frac{1}{4}AA + \frac{1}{2} + Aa + \frac{1}{4}aa$$
$$= \frac{1}{4}d + \frac{1}{2}h - \frac{1}{4}d = \frac{1}{2}h$$

Hence, the phenotypic value of $= \bar{F}_2 = m + \frac{1}{2}h$

For back cross generation *viz.*, B_1 and B_2 will have the phenotypic value

$$\bar{B}_1 = \frac{1}{2} + AA + \frac{1}{2}Aa$$
$$= \frac{1}{2}d + \frac{1}{2}h$$
$$= m + \frac{1}{2}d + \frac{1}{2}h$$
$$\bar{B}_2 = \frac{1}{2}aa + \frac{1}{2}Aa$$
$$= -\frac{1}{2}aa + \frac{1}{2}h$$
$$= m + \frac{1}{2}d + \frac{1}{2}h$$

Assuming several loci with additive effect $\Sigma d = (d)$ and $\Sigma h = (h)$

$$\bar{P}_1 = m + (d)$$
$$\bar{P}_2 = m - (d)$$
$$\bar{F}_1 = m + (h)$$
$$\bar{F}_2 = m + \frac{1}{2}(d)$$
$$\bar{B}_1 = m + \frac{1}{2}(d) + \frac{1}{2}(h)$$
$$\bar{B}_2 = m - \frac{1}{2}(d) + \frac{1}{2}(h)$$

With non– allelic interaction

A cross producing F_1, heterozygous for two gene pairs, can be made in two ways

(*i*) AABB × *aabb*

Genes to be associated AABB $= m + d_A + d_B + i_{AB}$; *aabb* $= m - d_A - d_B + i_{AB}$

$\bar{F}_1$ in both the crosses $= m + h_A + h_B + i_{AB}$

$\bar{F}_2$ in both the crosses $= m + \frac{1}{2}h_A + \frac{1}{2}h_B + \frac{1}{4}i_{AB}$ (in the absence of linkage)

$$\bar{B}_1 = m + \frac{1}{2}d_A + \frac{1}{2}d_B + \frac{1}{2}h_A + \frac{1}{2}h_B + \frac{1}{4}i_{AB} + \frac{1}{4}J_{AB} + \frac{1}{4}J_{BA} + \frac{1}{4}J_{AB}$$

$$\bar{B}_2 = m - \frac{1}{2}d_A - \frac{1}{2}d_B + \frac{1}{2}h_A + \frac{1}{2}h_B + \frac{1}{4}i_{AB} - \frac{1}{4}J_{AB} - \frac{1}{4}J_{BA} + \frac{1}{4}J_{AB}$$

(*ii*) AA*bb* × *aa*BB

Genes to be dispersed AAbb $= m + d_A - d_B - i_{AB}$; aaBB $= m - d_A + d_B + i_{AB}$

$\bar{F}_1$ in both the crosses $= m + h_A + h_B + i_{AB}$

$\bar{F}_2$ in both the crosses $= m + \frac{1}{2}h_A + \frac{1}{2}h_B + \frac{1}{4}i_{AB}$ (In the absence of linkage)

$$\bar{B}_1 = m + \frac{1}{2}d_A - \frac{1}{2}d_B + \frac{1}{2}h_A + \frac{1}{2}h_B - \frac{1}{4}i_{AB} - \frac{1}{4}J_{AB} - \frac{1}{4}J_{BA} + \frac{1}{4}J_{AB}$$

$$\bar{B}_2 = m - \frac{1}{2}d_A + \frac{1}{2}d_B + \frac{1}{2}h_A + \frac{1}{2}h_B - \frac{1}{4}i_{AB} - \frac{1}{4}J_{AB} - \frac{1}{4}J_{BA} + \frac{1}{4}J_{AB}$$

Using the generalized forms for [*d*], [*h*] and their interactions

$$\bar{P}_1 = m + [d] + [i]$$
$$\bar{P}_2 = m - [d] + [i]$$
$$\bar{F}_1 = m + [h] + [l]$$
$$\bar{F}_2 = m + \frac{1}{2}[h] + \frac{1}{4}[l]$$

$$\bar{B}_1 = m + \frac{1}{2}[d] + \frac{1}{2}[h] + \frac{1}{4}[i] + \frac{1}{4}[j] + \frac{1}{4}[l]$$

$$\bar{B}_2 = m - \frac{1}{2}[d] + \frac{1}{2}[h] + \frac{1}{4}[i] + \frac{1}{4}[j] + \frac{1}{4}[l]$$

LINKAGE

Polygene exhibit effect of linkage. Mean expressions of particularly of F_2 and subsequent generations are influenced by linkage. The effect of linkage on mean expression depends upon the strength of linkage *i.e.* p (recombination frequency).

Coupling Phase

$$\bar{F}_2 = m + \frac{1}{2}(h) + \frac{1}{2}(1 + 2p)(i) + \frac{1}{2}(1 + 2pq)\ (l)$$

Repulsion phase

$$\bar{F}_2 = m + \frac{1}{2}(h) - \frac{1}{2}(1 + 2p)(i) + \frac{1}{2}(1 + 2pq)\ (l)$$

Where, P = recombination frequency $(1 - q)$;

q = frequency of parental type $(1 - p)$

11 Variance Components of a Quantitative Trait

The genetic properties of a population are determined by the relative magnitudes of the components of variance. In addition, by knowing the components of variance, one may estimate the relative importance of the various determinants of phenotype. K. Mather expressed the phenotypic value of quantitative traits in this commonly used expression:

P (PHENOTYPE) = G(GENOTYPE) + E (ENVIRONMENT)

Individuals differ in phenotypic value. When the phenotypes of a quantitative trait are measured, the observed value represents the phenotypic value of the individual. The phenotypic value is variable because it depends on genetic differences among individuals, as well as environmental factors and the interaction between genotypes and the environment (called $G \times E$ interaction).

Total variance of a quantitative trait may be mathematically expressed as follows:

$$V_P = V_G + V_E + V_{GE}$$

where V_P = total phenotypic variance of the segregating population, V_G = genetic variance, V_E = environmental variance, and V_{GE} = variance associated with the genetic and environmental interaction.

The genetic component of variance may be further partitioned into three components as follows:

$$V_G = V_A + V_D + V_I$$

where V_A = additive variance (variance from additive gene effects), V_D = dominance variance (variance from dominance gene action), and V_I = **interaction** (variance from interaction between genes). Additive genetic variance (or simply additive variance) is the variance of breeding values and is the primary cause of resemblance between relatives. Hence V_A is the primary determinant of the observable genetic properties of the population, and of the response of the population to selection. Further, V_A is the only component that the researcher can most readily estimate from observations made on the population.

The total phenotypic variance may then be rewritten as:

$$V_P = V_A + V_D + V_I + V_E + V_{GE}$$

12 Heritability and Genetic Advance

Estimation of phenotypic and genotypic variances helps in the computation of heritability and genetic advance. Heritability is the proportion of the total variability that is due to genetic causes (or) it is the ratio of genotypic variance to the total variance. It is a good index of transmission of characters from parents to their offspring. Depending upon the components of variance used as numerator in the calculation of the heritability, it is of two types:

1. Broad sense heritability
2. Narrow sense heritability

The broad sense heritability is the ratio of genotypic variance to the total phenotypic variance and is calculated as

$$\text{Heritability } (H^2) = \frac{Vg}{VP} \times 100$$

Where H^2 = Heritability, Vg = Genotypic variance and Vp = Phenotypic variance

Narrow sense heritability is the ratio of additive (or) fixable genetic variance to the total phenotypic variance and is calculated as

$$h^2 = \frac{VA}{VP} \times 100$$

Heritability plays an important role in selection process in the plant breeding especially in the selection of elite genotypes from segregating populations. Johnson *et. al.*,(1995) categorized broad sense heritability as low (<30%) ; moderate (30–60%) and high (<60%),while, Robinson *et.al.* (1966) classified narrow sense heritability as low (5 to 10 %), medium (10–30 %) and high (> 30%).

FACTORS AFFECTING HERITABILITY ESTIMATES

The magnitude of heritability estimates depends on the genetic population used, the sample size, and the method of estimation.

1. **Genetic population:** When heritability is defined as h^2 = VA/VP (*i.e.*, in the narrow sense), the variances are those of individuals in the population. However, in plant breeding, certain traits such as yield are usually measured on a plot basis (not on individual plants). The amount of genotypic variance present for a trait in a population influences estimates of heritability. Parents are responsible for the genetic structure of the populations they produce. More divergent parents yield a population that is more genetically variable.

Inbreeding tends to increase the magnitude of genetic variance among individuals in the population. This means that estimates derived from F_2 will differ from, say, those from F_6.

2. **Sample size:** Because it is impractical to measure all individuals in a large population, heritabilities are estimated from sample data. To obtain the true genetic variance for a valid estimate of the true heritability of the trait, the sampling should be random. A weakness in heritability estimates stems from bias and lack of statistical precision.
3. **Method of computation:** Heritabilities are estimated by several methods that use different genetic populations and produce estimates that may vary. Common methods include the variance component method and parent–offspring regression. Mating schemes are carefully designed to enable the total genetic variance to be partitioned.

METHODS OF COMPUTATION

The different methods of estimating heritability have both strengths and weaknesses.

Evaluating Parental Germplasm

A useful application of heritability is in evaluating the germplasm assembled for a breeding project to determine if there is sufficient genetic variation for successful improvement to be pursued. A replicated trial of the available germplasm is conducted and analyzed by ANOVA as follows:

Sources	Degrees of freedom (df)	Expected mean sum of squares (EMS)
Replications	$r-1$	
Genotypes	$g-1$	$\sigma^2 + r\sigma^2_g$
Error	$(r-1)(g-1)$	σ^2

From the analysis, heritability may be calculated as:

Genotypic variance (σ^2_g) = Genotypic mean sum square – Error mean sum square / No. of replications

Error variance (σ^2_e) = Error mean sum square

Phenotypic variance (σ^2_p) = Genotypic variance (σ^2_g) + error variance (σ^2_e)

Heritability broad sense (H^2) = $\frac{Vg}{VP} \times 100$

It should be pointed out that whether the estimate is heritability in the narrow or broad sense depends on the nature of the genotypes. Pure lines or inbred lines would yield additive type of variance, making the estimate narrow sense. Segregating population would make the estimate broad sense.

Component of Variance Method:

This method is used to estimate narrow sense heritability. In this method phenotypic variance of F_2 and back crosses are partitioned into different components of D, H and E

When interaction is absent, then

$$VF_2 = \frac{1}{2}D + \frac{1}{4}H + E$$

$$VB_1 \text{ or } B_2 = \frac{1}{4}D + \frac{1}{4}H + E$$

$$h^2(F_2) = \frac{1}{2}D/VF_2$$

$$h^2(B_1) = \frac{1}{4}D/VB_1$$

$$h^2(F_2) = \frac{1}{4}D/VB_2$$

For F_3 data:

Variance of F_3 mean = $VF_3 = \frac{1}{2}D + \frac{1}{16}H + E = h^2 = \frac{1}{2}D/VF_3$

Mean Variance of F_3 line = $VF_3 = \frac{1}{4}D + \frac{1}{16}H + E = h^2 = \frac{1}{4}D/VF_3$

Covariance between F_2 parental mean and F_3 family mean

$\frac{1}{2}D + \frac{1}{8}H + E = h^2(ns) = \frac{1}{2}D/\text{Cov. } F_2\text{–}F_3$

APPROXIMATE ERROR VARIANCE METHOD

This method is used to estimate broad sense heritability in segregating generation.

Error variance (E)

1. $\frac{(VP_1 + VP_2)}{2}$ or $\frac{(VP_1 + VP_2)}{\frac{1}{2}}$

2. $\frac{(VP_1 + VP_2 + VF_1)}{3}$

For F_2 data

$$H^2(\%) = \frac{(VF_2 + VE)}{VF_2}$$

PARENT OFFSPRING REGRESSION METHOD

This method is used to estimate broad sense heritability only. The genetic portion of the total variance of population may be estimated by taking the regression of progeny means of the character on the parental value. The merit of this method is that it overcomes the variation due to seasonal effect and the extent of variation of the population has less impact on h^2 value. However, this method is not useful in resistance breeding.

Regression estimates for h^2

$$h^2 = 2b \text{ – For cross pollinated species}$$

$$h^2 = b \text{ – For self pollinated species}$$

In general case, as per Kempthorne (1957)

$$h^2 = \frac{b}{2rxy}$$

Where, b = regression value and rxy = Coefficient of parentage

As per Mahmud and Karmer (1951)

$$h^2 = \frac{h}{2rxy}$$

Where, $\bar{x}$ mean of parental generation; $\bar{y}$ = mean of later generation; by x = Regression of y on x

Applications of heritability

Heritability estimates are useful for breeding quantitative traits. The major applications of heritability are:

1. To determine whether a trait would benefit from breeding. If, in particular, the narrow sense heritability for a trait is high, it indicates that the use of plant breeding methods will likely be successful in improving the trait of interest.
2. To determine the most effective selection strategy to use in a breeding program. Breeding methods that use selection based on phenotype are effective when heritability is high for the trait of interest.
3. To predict gain from selection. Response to selection depends on heritability. A high heritability would likely result in high response to selection to advance the population in the desired direction of change.

Genetic Advance:

It is a measure of genetic gain under selection. It refers to the improvement in the mean genotypic value of the selected lines (or) families over the base population.

The expected genetic gain (or) advance under selection may be estimated Genetic advance is calculated by formula

Genetic advance (GA) = $k \times H^2 \times (V_P)^{1/2}$

where, k = Selection differential (2.06 at 5% selection intensity)

H^2 – Heritability (broad sense)

$(V_p)^{1/2}$ – Phenotypic standard deviation

$$\text{GA as a percentage of mean} = \frac{\text{GA}}{\text{Mean of the character}} \times 100$$

The success of G.A under selection depends

1. The magnitude of genetic variability present in base population
2. Heritability of a character under selection.
3. Intensity of selection

The range of genetic advance is classified by Johnson *et al* 1955 as low (<10 %); moderate (10–20%) and high (>20 %)

Heritability and genetic advance are important selection parameters and heritability estimate along with genetic advance are interpreted as follows

1. High heritability accompanied with high genetic advance indicates heritability is due to additive (or) fixable variation and selection may be effective.
2. High heritability accompanied with low genetic advance indicates non additive gene action and selection for such characters may not be rewarding.
3. Low heritability accompanied with high genetic advance reveals that characters are governed by fixable gene effects and low heritability is due to high environmental influence and selection may be effective.
4. Low heritability accompanied with low genetic advance indicates that character is highly influenced by environment and selection is ineffective.

SOLVED EXERCISE: ESTIMATION OF HERITABILITY AND GENETIC ADVANCE

Method–I : Evaluating Parental Germplasm

Twenty eight pearl millet genotypes have been evaluated in RBD with two replications. The data on green forage yield was analyzed and ANOVA for RBD is given below. Work out component of variance for green forage yield.

Analysis of variance for green forage yield (kg/plot)

Source	*df*	SS	MS	F cal	F table	
Replications	1	2.46	2.46	3.84	4.21	7.64
Treatments	27	736.91	27.29**	42.64	1.90	2.50
Error	27	17.23	0.64			
Total	55	756.60		G. Mean=5.42		

$$\text{Environmental variance} = \sigma^2_e = [\text{Error MS}] = 0.64$$

$$\text{Genotypic variance} = \sigma^2_g = \frac{[\text{MS genotypes} - \text{MS error}]}{r} = \frac{27.29 - 0.64}{2} = \mathbf{13.32}$$

$$\text{Phenotypic variance} = \sigma^2_p = [\sigma^2_g + \sigma^2_e] = 13.32 + 0.64 = 13.96$$

$$\text{Broad sense Heritability (\%)} = h^2 = \left(\frac{\sigma^2_g}{\sigma^2_p}\right) \times 100 = \left(\frac{13.32}{13.98}\right) \times 100 = 95.41$$

Genetic Advance = GA = $k \times h^2 \times \sigma_p = 2.06 \times 0.9541 \times (13.96)^{1/2} = 7.34$

Genetic advance as % of mean = $\left(\frac{GA}{GM}\right) \times 100 = \left(\frac{7.34}{5.42}\right) \times 100 = 135.42$

CONCLUSION

The estimates of both heritability and genetic advance as percent of mean are high. This indicated that the green forage yield of pearl millet is under the control of additive gene effect and hence selection will be effective.

SOLVED EXERCISE: HERITABILITY ESTIMATES BY COMPONENTS OF VARIANCE METHOD

An experiment was conducted with 6 generations (P_1, P_2, F_1, F_2, B_1, B_2) of pigeon pea cross. Following are the mean and variances of each generation for grain yield. Work out components of variances.

Generation	Mean	Variance
P_1	7.20	24.38
P_2	9.90	44.13
F_1	24.07	160.24
F_2	21.66	380.30
B_1	20.42	294.68
B_2	19.25	220.52

Estimation of components of variances, Additive (D), Dominance (D) and Environmental (E):

$$V F_2 = \frac{1}{2}D + \frac{1}{4}H + E \quad ...(1)$$

$$V B_1 \text{ or } V B_2 = \frac{1}{4}D + \frac{1}{4}H + E \quad ...(2)$$

i.e. $$2 V F_2 - (V B_1 + V B_2) = \frac{1}{2D}$$

$$V_E = \frac{(VP_1 + VP_2 + VF_1)}{3}$$

$$= \frac{(24.38 + 44.13 + 160.24)}{3}$$

$$= \frac{228.75}{3} = 76.25$$

$$2 V F_2 - (V B_1 + V B_2) = ½ D$$

$$= (2 \times 380.30) - (294.68 + 220.52)$$

$$= \frac{1}{2}D = (760.60 - 515.20) = \frac{1}{2}D$$

i.e.
$$\frac{1}{2}D = 245.40$$
$$D = 2 \times 245.340$$
$$= 490.80$$

Substitute the values of D and E in equation (I)

$$V_{F2} = \frac{1}{2}D + \frac{1}{4}H + E \quad \text{(I)}$$
$$= 380.80 = (0.5 \times 490.80) + \frac{1}{4}H + 76.25$$
$$\frac{1}{4}H = 380.30 - 245.40 - 76.25$$
$$= 4 \times (58.65) = 234.60$$

Narrow sense heritability (%) for $F_2 = [(\frac{1}{2}D)/V_{F2}] \times 100$

$$= \left(\frac{0.5 \times 490.80}{380.30}\right) \times 100 = 64.52\ \%$$

$$\text{Genetic advance} = k \times h^2 \times$$
$$= 2.06 \times 0.6452 \times 19.49$$
$$= 25.87$$

$$\text{Genetic advance as \% of mean} = \frac{\text{GA}}{\text{mean of } F_2 \text{ mean}} \times 100$$
$$= \frac{25.90}{21.66} \times 100$$
$$= 119.60$$

Estimation of heritability for backcross generations

$$VB_1 = \frac{1}{4}D + \frac{1}{4}H + E$$

$$294.68 = \left(\frac{1}{4} \times 490.80\right) + \frac{1}{4}H + 76.25$$

i.e.
$$1/4\ H = 294.68 - 122.70 - 76.25 = 95.73$$
$$H = 4 \times 95.73 = 382.92$$

Narrow sense heritability (%) for B_1

$$= \left[\frac{\left(\frac{1}{4}D\right)}{V_{B1}}\right] \times 100$$

$$= \left(\frac{0.25 \times 490.80}{294.68}\right) \times 100$$

$$= 41.64\ \%$$

$$\text{Genetic advance} = k \times h^2 \times$$

$$= 2.06 \times 0.4164 \times 17.17$$

$$= 14.73$$

$$\text{Genetic advance as \% of mean} = \left(\frac{\text{GA}}{\text{mean of B}_1}\right) \times 100$$

$$= \left(\frac{14.73}{20.42}\right) \times 100 = 72.13$$

$$V_{B2} = \frac{1}{4}\text{D} + \frac{1}{4}\text{H} + \text{E}$$

$$220.52 = \left(\frac{1}{4} \times 490.80\right) + \frac{1}{4}\text{H} + 76.25$$

i.e. $$\frac{1}{4}\text{H} = 220.52 - 122.70 - 76.25 = 21.57$$

$$\text{H} = 4 \times 21.57 = 86.28$$

$$\text{Narrow sense heritability (\%) for B}_2 = \left[\frac{\left(\frac{1}{4}\text{D}\right)}{V_{B2}}\right] \times 100$$

$$= \left(\frac{0.25 \times 490.80}{220.52}\right) \times 100$$

$$= 55.64\ \%$$

$$\text{Genetic advance} = k \times h^2$$

$$= 2.06 \times 0.5564 \times 14.85$$

$$= 17.02$$

$$\text{Genetic advance as \% of mean} = \left(\frac{\text{GA}}{\text{mean of B}_2}\right) \times 100$$

$$= (17.02/19.25) \times 100$$

$$= 88.41$$

CONCLUSION

The estimates of narrow sense heritability and genetic advance as % mean were also higher in magnitude indicating that the character is predominantly under the control of additive gene action and hence selection in segregating generations will be effective for improvement of grain yield of pigeon pea.

SOLVED EXERCISE : ESTIMATION OF HERITABILITY BY PARENT OFFSPRING REGRESSION ANALYSIS

The parent value of 30 (F_2) and its corresponding progeny (F_3) means for yield/ plant are given below. Estimate regression coefficient and heritability.

Sr.No	F_2	F_3	$(F_2 \times F_3)$	$(F_2)^2$	Sr.No	F_2	F_3	$(F_2 \times F_3)$	$(F_2)^2$
1	33.66	59.32	1996.71	1133.00	17	33.30	67.30	2241.09	1108.89
2	34.00	58.30	1982.20	1156.00	18	36.70	65.11	2389.54	1346.89
3	31.10	38.97	1211.97	967.21	19	27.70	46.70	1293.59	767.29
4	28.80	44.00	1267.20	829.44	20	26.70	47.50	1268.25	712.89
5	27.20	43.30	1177.76	739.84	21	37.11	48.20	1788.70	1377.15
6	31.12	49.08	1527.37	968.45	22	37.70	46.80	1764.36	1421.29
7	28.60	48.00	1372.80	817.96	23	40.80	62.32	2542.66	1664.64
8	33.00	73.26	2417.58	1089.00	24	39.50	60.60	2393.70	1560.25
9	32.10	72.20	2317.62	1030.41	25	38.30	60.30	2309.49	1466.89
10	30.40	70.40	2140.16	924.16	26	30.40	44.80	1361.92	924.16
11	34.10	42.11	1435.95	1162.81	27	28.30	45.60	1290.48	800.89
12	30.50	46.30	1412.15	930.25	28	31.70	52.16	1653.47	1004.89
13	29.40	47.40	1393.56	864.36	29	31.50	50.40	1587.60	992.25
14	32.20	53.54	1723.99	1036.84	30	41.30	50.00	2065.00	1705.69
15	30.10	50.30	1514.03	906.01	Total	**981.29**	**1612.38**	**53156.63**	**32565.80**
16	34.00	68.11	2315.74	1156.00	Mean	32.71	53.75		

$$\text{Regression of } F_3 \text{ on } F_2 = \frac{\text{Sum of product } F_3\text{–}F_2}{\text{Sum of square of } F_2}$$

$$= [\text{sum of product of corres. Value of F2 \& F3}] - \left(\frac{\text{GT of F2} \times \text{GT of F3}}{N_2}\right)$$

$$\text{SP of } F_3\text{–}F_2 = (33.66 \times 59.32) + ... + (41.30 \times 50.0) - \left(\frac{981.29 \times 1612.38}{30}\right)$$

$$= 53156.63 - 52740.41 = 416.22$$

SS of F_2 = sum of squares of Value of F_2 – (GT of F_2)^2/No. of F_2 plants /N

$$= \text{Sum of } (33.66)^2 + ... + (41.30)^2 - \frac{(981.29)^2}{30}$$

$$= 32565.80 - 32097.6688 = 468.13$$

$$\text{Regression of } F_3 \text{ on } F_2 = \frac{\text{Sum of product } F_3\text{–}F_2}{\text{Sum of square of } F_2}$$

$$\frac{416.22}{468.13} = 0.89$$

$$h^2 = \frac{2}{3b} = 0.67 \times 0.89 = 0.59 \text{ i.e., } 59\ \%$$

$$h^2 = \left(\frac{\text{Mean } F_2 \times b\ F_3 \text{ on } F_2}{\text{Mean of } F_3}\right) \times 100$$

$$= \frac{(32.17 \times 0.89)}{53.75} \times 100$$

$$= \left(\frac{29.11}{53.75}\right) \times 100 = 54.16\ \%$$

CONCLUSION

The parent progeny regression (*b*) value is high and heritability estimated is also moderate .

Realized Heritability:

The heritability which is estimated by selection experiments on the basis of response to selection

$$R = h^2S \qquad h^2 = R/S$$

Where, S = Selection differential and h^2 = heritability

Selection Differential: Selection differential is defined as the difference between the mean phenotypic value of the individuals selected to be the parents and whole of the initial generation before selection.

Repeatability

The measure of association among the repeated records on the same individual over a period of time is called repeatability. In statistical term, it is intra class correlation among the repeated records on the same individual over a period of time. Repeatability is estimated for the characters which are repeated over the time (lactation yield, fruit yield etc).

It is represented by (*r*) by Lush (1937)

$$\text{Repeatability } (r) = \frac{Vg + VEg}{Vp}$$

Where, Vg = Genotypic variance; Vp = Phenotypic variance and VEg = General environmental variance which is due to permanent on non– localized environmental situation.

Utility: It sets the upper limit of broad sense heritability. The approx. estimate of repeatability is used to estimate broad sense heritability when it is not possible to estimate broad sense heritability and it also indicates the extents to which early performance can be taken as a indicator of later performance.

13

Correlated Response

Correlation is a measure of the degree of association between traits. This association may be on the basis of genetics or may be non–genetic. In terms of response to selection, genetic correlation is what is useful. When it exists, selection for one trait will cause a corresponding change in other traits that are correlated. This response to change by genetic association is called correlated response. Correlated response may be caused by pleotropism or linkage disequilibrium. Pleotropism is the multiple effect of a single gene (*i.e.*, a single gene simultaneously affects several physiological pathways). In a random mating population, the role of linkage disequilibrium in correlated response is only important if the traits of interest are closely linked.

In calculating correlated response, genetic correlations should be used. The correlated response (CR*y*) to selection in the primary trait (*y*) for a secondary trait (*x*) is given by:

$$CRy = i_x h_x h_y \varrho_g \sqrt{V_{Py}}$$

where h_x and h_y are square roots of the heritabilities of the two respective traits, ϱ_g is the genetic correlation between traits and i_x is selection intensity. This relationship may be reduced to:

$$CRy = i_x \varrho_g h_x \sqrt{V_{Gy}} \text{ since } h_y = \sqrt{(V_{Gy}/V_{Py})}$$

It is clear that the effectiveness of indirect selection depends on the magnitude of genetic correlation and the heritability of the secondary traits being selected.

Further, given the same selection intensity and a high genetic correlation between the traits, indirect selection for the primary trait will be more effective than directional selection, if heritability of the secondary trait is high ($\varrho_g h_x > h_y$). Correlated response has wider breeding application in homozygous, self–fertilizing species and apomicts.

Additive genetic correlation is important in selection in plant breeding. The additive breeding value is what is transferred to offspring and can be changed by selection. Hence, where traits are additively genetically correlated, selection for one trait will produce a correlated response in another.

14 Discriminant Function and Selection Index

It is a statistical approach being used in construction of Selection index and discriminate desirable genotypes from undesirable ones based on various characters associated with yield.

Fisher (1936) has given this function with a purpose to discriminate individuals which belong to two different populations showing some degree of overlapping and mathematically it can be defined as

$$Z = b_1x_1 + b_2x_2 + b_3x_3 ... + b_nx_n$$

Where

Z = Discriminant function

$x_1, x_2, x_3 ... x_n$ are measured variables

$b_1, b_2, b_3, ... b_n$ are weighing coefficients.

The b_i values are estimated such that they are based on Z values. The ratio of variance between populations would be maximized the same maximized ratio leads to a set of simultaneous equations which after solution provide the desired values of b_i.

MAIN FEATURES OF DISCRIMINATE FUNCTION ANALYSIS

(1) It measures the efficiency of various characters combinations in selection. Selection index leads to simultaneous manipulations of several characters for genetic improvement of economic yield.

(2) This technique provides information on yield components and thus helps in indirect selection for genetic improvement of yield.

(3) Analysis is based on linearity and additivity.

(4) Analysis involves variance and covariance.

(5) Though based on path analysis it differs from it in several aspects.

Difference between Path Coefficient Analysis and Discriminate function

Path Coefficient Analysis	Discriminate function
It measures the cause of association between two variables.	It measures the efficiency of various traits combinations in selection.
Analysis is based on all possible simple correlations.	Analysis involves variances and co–variances.
Direct, indirect and residual effects are calculated for path analysis.	Weight coefficients, expected genetic advance and relative efficiency are estimated.

Similarities:

(1) Both help in determining the yield components.

(2) Both based on statistical assumptions of linearity and additivity.

Selection index: Selection index can be defined as linear combinations of plant characters associated with yield. Smith (1936) first proposed the discriminant function of Fisher for plant selection. The best known selection indices involve discriminant function based on the relative economic importance of various characters. Selection indices or index are three types, viz., classical, general and restricted.

(*i*) **General selection index:** Hanson and Johnson (1957) first proposed general selection index. The construction of general selection index is based on pooled information of different populations from which b_i values are estimated. Using these b_i values the genotypic and phenotypic variances /co–variances are corrected then these corrected variances/co–variances are used to generate new b_i values. These b_i values can be applied to any population.

(*ii*) **Classical selection index:** Smith (1936) proposed a classical selection index model The application of discriminate function as a basis for making selection on several characters simultaneously is aimed at discriminating the desired genotypes from undesired ones on the basis of their phenotypic performance. According to selection indices involved two functions

(1) The genetic worth of an individual 'H' and

(2) The Phenotypic performance of various characters of plant 'I' . The coefficient of the function I are estimated such that the correlation between H and I *i.e.* r (H, I) is maximized. This leads to a set of simultaneous equations which upon solving give desired estimates of b_i values. These indices are specific to those populations from which they have been derived.

(*iii*) **Restricted selection index:**This selection index was proposed by Kempthorne and Nordskog (1959). This helps in improving a set of characters keeping the value of other character intact. Sometimes the restriction is put on single and sometimes on double characters.

COMPUTATION OF SELECTION INDEX

Selection index can be constructed using one of the three models described above. Construction of selection index is based on the estimates of phenotypic and genotypic variances and co–variances of characters involved in selection index. The selection indices can be worked out from replicated data only. Classical selection index model proposed by Smith (1936) is more effective than other models.

CLASSICAL SELECTION INDEX MODEL

The genotypic effect of all component effects can be represented by a function

$$H = a_1G_1 + a_2G_2 + a_3G_3 + \dots\dots\dots\dots a_nG_n$$

Where

$G_1, G_2 \ldots G_n$ are the genotypic values of individual characters and $a_1, a_2 \ldots a_n$ are weights depending upon the relative importance in the contribution of component characters towards the composite character.

Weighing Coefficients

The coefficients refer to the relative importance of various characters in the index. The general form of selection index is given below

$$I = b_1x_1 + b_2x_2 + b_3x_3 + \ldots\ldots\ldots\ldots\ldots\ldots b_nx_n$$

where $x_1, x_2, x_3 \ldots\ldots x_n$ represent the phenotypic values of the character number 1,2,3, and n respectively and $b_1, b_2, b_3 \ldots b_n$ corresponding weights. $b_1, b_2, b_3 \ldots b_n$ are to be estimated such that the correlation between H and I becomes maximum.

The b values are calculated from series of simultaneous equations involving the appropriate phenotypic and genotypic variances and covariance. The b values are worked out separately for various selection indices involving single, double, triple and multiple traits. The simultaneous equations are solved by elimination process and b values are obtained. If there were three characters and yield the following simultaneous equations could be set with the help of appropriate variances and covariance of these traits.

$$b_1x_{11} + b_2x_{12} + b_3x_{13} = a_1G_{11} + a_2G_{12} + a_3G_{13}$$
$$b_1x_{21} + b_2x_{22} + b_3x_{23} = a_1G_{21} + a_2G_{22} + a_3G_{23}$$
$$b_1x_{31} + b_2x_{32} + b_3x_{33} = a_1G_{31} + a_2G_{32} + a_3G_{33}$$

Which in matrix form is

$$\underset{A}{\begin{pmatrix} x_{11} & x_{12} & x_{13} \\ x_{21} & x_{22} & x_{23} \\ x_{31} & x_{32} & x_{33} \end{pmatrix}} \times \underset{B}{\begin{pmatrix} b_1 \\ b_2 \\ b_3 \end{pmatrix}} = \underset{C}{\begin{pmatrix} G_{11} & G_{12} & G_{13} \\ G_{21} & G_{22} & G_{23} \\ G_{31} & G_{32} & G_{33} \end{pmatrix}} \underset{D}{\begin{pmatrix} a_1 \\ a_2 \\ a_3 \end{pmatrix}}$$

The solution of these equations give the estimates of bi values;

B = $A^{-1}CD$

A = Phenotypic variance and co–variance matrix

C = Genotypic variance and co–variance matrix

D = Weights

Assign weights for phenotypic value

(a) Equal weights: Assume all the characters are economically importance *i.e.*, $a_1 = a_2 = a_3 = 1$

Equal weights can be obtained by reciprocal of means, as the mean of different characters may have biasness of unit of measurement.

$$a_1 = \frac{1}{x_1}; a_2 = \frac{1}{x_2}; a_3 = \frac{1}{x_3}$$

Relative weights: Assign weights depending upon importance of the characters. If the relative importance of x_1, x_2 and x_3 are 3:2:1, then $a_1 = \frac{3}{x_1}; a_2 = \frac{2}{x_2}; a_3 = \frac{1}{x_3}$

On the basis of selection index, the genotypes are arranged in the order of merit and the the topranking 6 or 10% may be selected for further breeding programme.

(ii) **Expected Genetic Gain:** The genetic gain through selection may be predicted with the help of following formula

$$\text{Genetic Gain} = K \frac{W}{(VP)^{1/2}}$$

$$\Delta G = \frac{Z}{Q}(b_1 g_1 y + b_2 g_2 y + \ldots . b_n g_n y)^{1/2}$$

Where,

k = selection differential for 5 % or 10 %

$W = \Sigma\Sigma\, a_i b_j G_{ij}$

ai = Weightage

bj = Regression coefficient

G_{ij} = Genotypic variance and co–variance matrix

$Vp = \Sigma\Sigma\, b_i b_j P_{ij}$

P_{ij} = Phenotypic variance and co–variance matrix

RELATIVE EFFICIENCY

The relative efficiency of selection is worked out separately for various selection indices involving single, double, triple and multiple traits with the help of formula

$$\text{Relative efficiency} = \frac{GS_1}{GS_2} \times 100$$

Where,

GS_1 – Discriminate function and GS_2 – Genetic advance for direct selection

Thus apparently it is expressed as the percentage ratio of genetic advance for discriminate function (GS_1) to genetic advance for direct selection (GS_2). It measures the effectiveness of various selection indices. The relative efficiency of direct selection for yield is considered as 100%. Any selection index showing superiority over direct selection is considered important and the characters involving in such index are considered as major components of yield. Such combination of characters has to be given due weightage in selection of elite genotypes from diverse breeding populations.

Analysis of selection indices are to be carried out according to the procedure given by Singh and Choudhary (1985).

15 Metroglyph Analysis

Metroglyph technique is proposed by Anderson (1957) to study the pattern of morphological variation in crop species. It is semi graphic method for assessing the variability in a large number of genotypes. The analysis is based on first order statistics and hence results are more reliable. Analysis can be carried out from both replicated as well as un replicated data.

STEPS IN THE ANALYSIS

1. **Selection of genotypes :** The material which is used for the assessment of variability may include germplasm lines, strains, varieties and hybrids.
2. **Evaluation of material :** The selected genotypes are tested in replicated or un replicated trials and observations are recorded on various characters and means are worked out.
3. **Assessment of variability:**
   - (*i*) Two most variable characters are selected. One of them is used on the X-axis and other on Y–axis. Then mean values of X for each genotype are plotted on the graph against the mean values of Y. For plotting the position of genotype in the graph, the index score of 1, 2 and 3 are used. Thus each genotype occupies a definite position on the graph called glyph. Variation for remaining characters of each genotype is displayed on the respective glyph by rays. Each character occupies a definite ray position. .
   - (*ii*) Variation for each character is depicted by the length of ray. Thus, the length of ray for a particular character on the glyph may be short (1) or medium (2) or long (3) depending on the index score values of the genotype.
   - (*iii*) For plotting of glyph on the graph, the X and Y axis of the graph are divided into three parts *viz.*, low (1), medium (2) and high (3). Thus, there will be nine parts of the clusters in the graph.
   - (*iv*) The range of variability in each quantitative character is also classified into three groups viz., low, medium and high. These groups are given an index score of 1, 2 and 3, respectively.

SOLVED EXERCISE

The following is a mean data 12 genotypes of cotton and the scores (in brackets) of for four characters. Construct the index score and group the genotypes as per metroglyph analysis suggested by Anderson (1957).

Genotypes	No of Sympodia/ plant	Boll weight (g)	No. of bolls/ plant	Seed cotton yield (g/ plant)	Total index score
1	11.07 (1)	3.03 (1)	41.40 (3)	106.93 (3)	2
2	16.27 (3)	3.44 (2)	35.07 (3)	67.67 (1)	5
3	15.40 (2)	3.28 (1)	21.47 (1)	72.42 (2)	3
4	14.80 (2)	3.71 (3)	24.07 (1)	53.97 (1)	5
5	13.80 (2)	3.57 (2)	34.20 (2)	91.00 (3)	4
6	13.00 (1)	3.49 (2)	32.20 (2)	87.53 (2)	3
7	13.13 (1)	3.88 (3)	30.27 (2)	80.02 (2)	4
8	12.20 (1)	3.06 (1)	26.47 (1)	70.20 (1)	2
9	17.80 (3)	3.48 (2)	35.53 (3)	86.73 (2)	5
10	13.53 (2)	3.62 (3)	31.60 (2)	100.32 (3)	5
11	13.73 (2)	3.79 (3)	28.47 (2)	52.95 (1)	5
12	12.33 (1)	3.02 (1)	20.73 (1)	54.58 (1)	2
Range	11.07–17.80	3.02–3.88	20.73– 41.40	52.95–106.53	

Figures in parenthesis are index score of the genotype. (Total score of characters excluding those used for X and Y axis)

ESTIMATION OF INDEX SCORES

Using suitable class intervals, the range of variability to a character is classified into three groups. From example, No. of sympodia the minimum range is 11.07 and maximum range is 17.80. The difference is 6.73. Divide this difference by 3 (No. of classes). The value is 2.24. Add this value in minimum value *i.e.,* 11.07 +2.24 = 13.31. In all the cases where no. of sympodia < 13.31, index score given is 1. Similarly, subtract this value from maximum value *i.e.* 17.80 – 2.24 = 15.56. Then individuals between 13.31 to 15.56 get and index value of 2 and those having sympodia > 15.56 assigned to an index value of 3. Similarly, for all other characters such scoring is done as given below.

Character	Range			Score 1		Score 2		Score 3	
	Min	Max	Diff./3	Value < than	Sign	Value from to	Sign	Value > than	Sign
No. of Sympodia	11.07	17.80	2.24	13.31	0	13.31 to 15.56	0	15.56	0
Boll weight (g)	3.02	3.88	0.29	3.31	0	3.31 to 3.59	0	3.59	0

SCATTER DIAGRAM IN METROGLYPH ANALYSIS

Let us plot Y values (mean No. of bolls/plant) against X values (Mean seed cotton yield) for each individual on graph. A glyph *i.e.* particular point (represented by open circle) obtained by plotting Y values(41.40) against X value (say, 106.93). On each specific position is fixed for a particular character. The performance of genotype is thus denoted by the index score of that genotype.

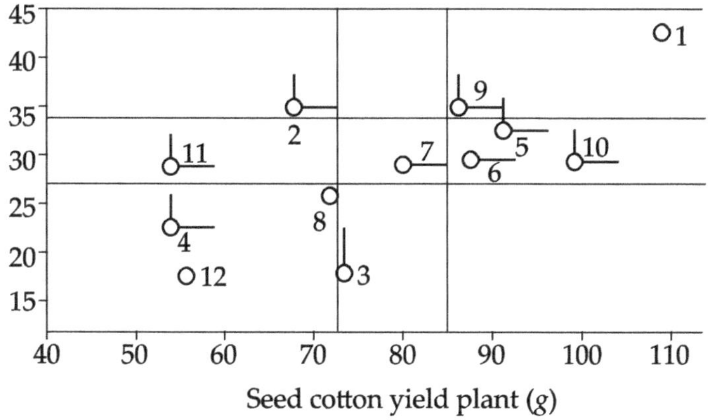

CONCLUSION

On the basis of wide variability for seed cotton yield and number of bolls, the genotype 1 ranked first followed by 10 and 5. The genotypes 6 and 7 groped into medium score group. Hence, on the basis of mean performance and total index score, the genotypes *viz.*, 1, 10, 5, 9, 7 and 2 may be used for crossing programme.

16

D^2 Analysis

The concept of D^2 statistic originally developed by P. C. Mahalanobis in 1928 to study anthropometry and psychometry. Rao (1952) suggested the application of this technique for the assessment of genetic diversity in plant breeding.

In plant breeding, genetic diversity plays an important role because hybrids between lines of diverse origin, generally, display greater heterosis than those between closely related parents. Genetic diversity arise due to geographical separation or due to genetic barriers to crossability.

Main features of D^2 analysis are:

1. This is a numerical approach
2. Estimates of D^2 statistics are based on second order statistics
3. Analysis is more difficult
4. Analysis is possible from replicated data only
5. Genetic diversity is depicted by cluster diagram

Main steps for analysis:

1. **Selection of genotypes**: Genotypes *i.e.* germplasm lines, strains or varieties are generally selected on the basis of phenotypic variability or geographical origin.
2. **Evaluation of material** : Selected genotypes are evaluated in replicated trial and observations are recorded on various characters.
3. **Biometrical Analysis:** From the data of replicated trial, genotypic and error variances for all the characters are calculated. From these variances and covariances, two matrices viz., error vaiance and covariance matrix and errot + genotypic variance and covariance matrix is formed.

   (*i*) **Test of significance**: V statistics (Wilk's criterion) which is the simultaneous test of differences between mean values of number of correlated variables.

   It is estimated using the relationship

$$\Lambda = \frac{|E|}{|E + V|}$$

Where IEI is the determinant of the experimental error sum of squares and sum of products matrix and $|E + V|$ the determinant of experimental error sum of squares and sum of products, plus the population sum of squares and product matrix significance of was estimated by X^2 as,

$$X^2\,pq = V = -m \log_e \Lambda$$

$$m = n - \frac{p+q+1}{2}$$

where,

n = degree of freedom for error + genotype

= Number of characters

q = degree of freedom for genotype

e = 2.718

$\log_e \quad \Lambda = 2.3026 \log_{10} \Lambda \qquad i.e.\ V_{(Stat)} = -m\,(2.3026) \log_{10} \Lambda$

The $V_{(stat)}$ value is compared with X^2 table value for *pq df*. The significance of $V_{(stat)}$ shows that the differences between the means in respect of the pooled effect of 'P' characters between genotypes are significant. Hence, further analysis for estimation of D^2 value can be proceeded.

(ii) Mahalanobis generalized distance (D^2)

The generalized distance between any two populations is defined as :

$$D^2 = \Sigma\Sigma \lambda ij \delta i \delta j$$

where,

λ_{ij} = Reciprocal matrix to the common dispersion matrix

δ_i = Difference between the mean values of the two populations for the ith character.

This quantity is estimated by the D^2 statistics as : $D^2 = \Sigma\Sigma Sij d_i d_j$

Where, Sij is the sample estimate of λij and di of δi. The formula for computation requires the inversion of the determinant and then evaluation of $n(n + 1/2)$ term whose sum is D^2.

COMPUTATION OF D^2 VALUES

For each combination, mean deviation, *i.e.* $Y_i^1 - Y_i^2$ with $i = 1, 2 \ldots\ldots p$, is computed and the D^2 is calculated as sum of the squares of these deviations *i.e.* $\Sigma(Y_i^1 - Y_i^2)^2$.

Determination of Population Constellations

No rules can be laid down for finding the clusters because cluster is not a well defined term. The only criterion appears to be that, any two genotypes belonging to the same clusters should be at least, on an average show a smaller D^2 values than those belonging to two different clusters.

(iii) Grouping of genotypes into clusters:

A simple method, suggested by Tocher (Rao, 1952) for cluster formation is to start with two closely related genotypes and find a third genotype which has the smaller average D^2 value from the first two. Similarly, the fourth genotype is chosen to have the smaller average D^2 value from the first three and so on. While proceeding further for the cluster formation, if at any stage, the average D^2 value of a genotype appears to be high than those already listed, then this genotype does not fit in that former group and taken to be outside of that cluster. The varieties included in the first cluster, are then omitted and the rests are treated similarly to form next cluster.

(*iv*) **Average intra cluster distance :** The intra cluster distances are calculated as $\frac{\Sigma Di^2}{n}$, where ΣDi^2 is the sum of distance between all possible combinations of (n) genotypes included in a cluster.

n = Number of genotypes included in cluster

(*v*) **Average inter cluster distances:** The procedure followed for calculating the inter cluster distances is first to measure the distance between the various combinations of clusters and devide by the product of number of genotypes in the concerned cluster combinations.

The average inter cluster distances are then calculated as : $\frac{\Sigma D_i^2}{(n_i n_j)}$

where,

n_i = Number of genotypes in cluster i

n_j = Number of genotypes in cluster j

The intra and inter cluster distance (D) values are then obtained by taking square root of average D^2 values of the respective genotypes.

(*vi*) **Cluster diagram :** With the help of D^2 values between (inter–cluster distance) and within (intra–cluster distance) clusters, a diagram showing the relationships between different genotypes can be drawn.

(*vii*) **Contribution of individual characters towards total divergence:** In all combinations of genotypes, (n(n–1)/2, each character is ranked on the basis of values ($Yi^1 - Yi^2$). Rank first is given to the highest mean difference and rank p to the lowest mean difference, where p is the total number of characters. Using these ranks, percent contribution of each character to the total divergence is estimated.

$$\text{Percent contribution by } Xi = \frac{\text{No. of times appearing first in rank by } x_1}{n(n-1)/2} \times 100$$

Where,

$Xi...X_p$ are individual characters n is the number of genotypes.

EXERCISE AND INTERPRETATION OF RESULTS

A trial was conducted with 28 genotypes of napier in a RBD with two replications. The data on eight quantitative characters were measured and D^2 analysis was performed by using Windostat software.

Mean performance of the genotypes for eight characters

S.N.	Gen.	Plant height	No. of tillers	No. of leaves	Inter-node length	Leaf length	Leaf width	DMY (kg)	Forage yield(kg)
1	G–1	146.70	12.73	10.32	7.67	63.78	2.38	0.87	3.33
2	G–2	142.09	8.08	11.77	6.89	77.11	2.42	0.76	3.15
3	G–3	153.14	6.41	12.50	7.29	80.35	2.43	0.63	2.87
4	G–4	161.20	10.08	11.81	7.22	74.77	2.37	0.99	3.63
5	G–5	143.73	5.90	11.04	5.97	85.35	2.62	0.61	2.81
6	G–6	149.42	8.60	11.72	6.88	81.33	2.60	0.79	4.01
7	G–7	169.30	9.87	12.67	7.37	87.08	2.83	1.34	5.67
8	G–8	169.44	13.02	12.12	7.57	64.12	2.80	1.89	6.61
9	G–9	105.87	9.33	11.78	4.85	58.95	2.39	0.57	2.44
10	G–10	101.07	8.87	11.87	4.57	60.97	2.25	0.44	1.83
11	G–11	169.99	15.25	11.62	6.92	79.94	2.32	1.04	3.65
12	G–12	150.02	10.50	10.78	6.90	66.30	2.54	0.60	2.05
13	G–13	146.47	9.82	10.77	6.59	67.09	1.96	0.61	2.06
14	G–14	123.94	10.13	11.10	5.85	62.65	1.81	0.44	1.61
15	G–15	135.39	10.25	10.93	6.37	71.35	1.92	0.62	3.01
16	G–16	172.29	8.58	12.93	7.34	83.72	2.33	1.25	5.70
17	G–17	166.89	22.52	13.10	9.40	63.22	2.06	1.25	4.65
18	G–18	237.50	16.18	14.10	10.29	77.92	2.00	1.75	6.08
19	G–19	218.80	16.33	14.27	10.84	70.53	1.89	2.07	7.75
20	G–20	200.40	13.27	13.38	9.94	77.19	1.88	1.39	5.08
21	G–21	214.84	29.90	14.34	11.10	86.44	3.66	4.50	17.64
22	G–22	176.92	12.45	12.80	8.54	76.10	2.60	1.20	5.23
23	G–23	190.83	6.75	12.25	8.25	78.59	2.68	0.71	2.66
24	G–24	232.48	23.72	15.25	12.25	86.92	2.48	2.44	8.87
25	G–25	220.40	16.70	15.33	10.90	83.53	3.60	2.16	8.69
26	G–26	235.87	20.00	12.99	10.28	90.57	2.71	2.14	7.17
27	G–27	224.75	22.97	11.84	8.25	95.17	3.00	3.03	11.49
28	G–28	205.20	26.12	12.79	8.23	93.47	3.29	2.12	12.13
	Mean	173.75	13.73	12.43	8.02	76.59	2.49	1.36	5.42

Estimation of V Statistics (Wilkinson Criteria)

V statistics = 674.62**

The significance of V stat. and MS due to varieties indicated that there were significant differences between 28 genotypes for the pooled effect of the eight characters tested. Hence, further analysis was done to calculate D^2 values.

Grouping of Genotypes into Clusters (Tocher's Method)

From the estimated D^2 values for all pair combinations, total 28 genotypes were grouped into following four clusters. The cluster I was the largest comprising 21 genotypes, while the clusters III and IV had 3 genotypes each, whereas cluster–II was solitary. Clustering pattern obtained revealed that the genetic diversity was not necessarily parallel to the geographical diversity. This means the genotypes originating in different geographical area could formed one cluster, while different genotypes evolved in the same area were grouped into different clusters.

Cl. No.	No. of gen.	Genotype	Origin	Genotype	Origin
I	21	G–09	Rahuri	G–11	Rahuri
		G–10	Rahuri	G–22	Coimbatore
		G–14	Rahuri	G–01	Rahuri
		G–13	Rahuri	G–17	Rahuri
		G–15	Rahuri	G–23	Hybreradad
		G–02	Rahuri	G–07	Rahuri
		G–03	Rahuri	G–20	Coimbatore
		G–04	Rahuri	G–16	Rahuri
		G–06	Rahuri	G–08	Rahuri
		G–05	Rahuri	G–24	Hybreradad
		G–12	Rahuri		
II	1	G–26	Hybreradad		
III	3	G–18	Rahuri		
		G–19	Rahuri		
		G–25	Hybreradad		
IV	3	G–27	Hybreradad		
		G–28	Hybreradad		
		G–21	Coimbtore		

Relative contribution of individual character to the total genetic divergence

Source	Times Ranked 1st	Contribution %
Plant height (cm)	102	26.98%
No. of tillers/plant	7	1.85%

Contd...

Contd...

No. of leaves/tiller	13	3.44%
Length of internodes (cm)	20	5.29%
Leaf length (cm)	56	14.81%
Leaf width (cm)	45	11.90%
Dry matter yield /plant	3	0.79%
Green forage yield/plant	132	34.92%

The maximum contribution to the total genetic divergence was due to the green forage yield (34.92 %) followed by plant height (26.98%), leaf length (14.81%) and leaf width (11.90%).

Average intra–cluster and inter–cluster distances (Digonal intra cluster and above diagonal inter cluster distances)

	Cluster–I	**Cluster–II**	**Cluster–III**	**Cluster–IV**
Cluster–I	**8.55**	12.53	13.41	23.78
Cluster–II		**0.00**	8.94	17.25
Cluster–III			**9.59**	19.90
Cluster–IV				**11.53**

An examination of intra–cluster divergence among four clusters revealed that, cluster IV had maximum intra–cluster distance (11.53) followed by Cluster–III (9.59) and Cluster–I (8.55). Whereas, the maximum inter cluster distance was found between the cluster I and cluster–IV (23.78) followed by cluster– III and cluster–IV (19.90) and cluster–II and cluster–IV (17.25).

Cluster Means for eight characters

Cluster No.	**Plant height**	**No. of tillers**	**No. of leaves**	**Internode length**	**Leaf length**	**Leaf width**	**Dry matter yield**	**Green forage yield**
I	157.50	11.24	12.02	7.36	73.66	2.36	0.97	3.85
II	**235.87**	20.00	12.99	10.28	90.57	2.71	2.14	7.17
III	225.57	16.40	**14.57**	**10.67**	77.33	2.50	1.99	7.51
IV	214.93	**26.33**	12.99	9.19	**91.69**	**3.32**	3.21	13.75

Cluster diagram:

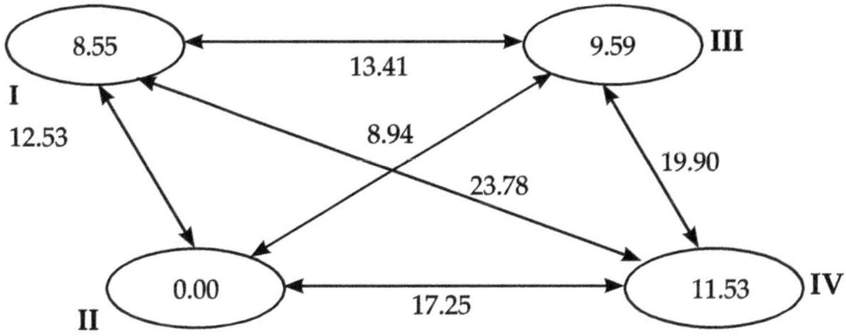

The data on cluster mean indicated that the genotypes grouped in cluster –IV had high mean for green forage yield, dry matter yield, leaf length, leaf width and no.of tillers; genotypes grouped in cluster–III had higher no. of leaves and internode length, whereas genotypes in cluster–II possessed maximum plant height.

CONCLUSION

On the basis of relative contribution of character to the total genetic divergence, inter cluster distance and cluster means the genotypes Viz., G–7, G–24, G–16 (Cluster–I); G–26 (Cluster–II; G–19 and G–25 (Cluster–III) ; G–21 and G–27 (Cluster–IV) are selected and suggested for further hybridization programme in diallel mating design.

17 Principal Component Analysis

Principal component analysis (PCA) was invented in 1901 by Karl Pearson, as an analogue of the principal axes theorem in mechanics; it was later independently developed named by Harold Hotelling in the 1930s. PCA is a statistical procedure that uses an orthogonal to convert a set of observations of possibly correlated variables into a set of values of linearly uncorrelated variables called principal components. It is a sort of multivariate analysis where canonical vectors or roots representing different axes of differentiation and the amount of variation accounted for each of such axes are derived. In other words, Principal component analysis is a mathematical procedure that transforms a number of correlated variables into a smaller number of uncorrelated variables called principal components.

SALIENT FEATURES OF PRINCIPAL COMPONENT ANALYSIS (PCA)

1. It is a variable reduction technique.
2. It is used when variables are highly correlated.
3. It reduces the number of observed variables to a smaller number of principal components which account for most of the variance of the observed variables.
4. It is a large sample procedure.
5. The total amount of variance in PCA is equal to the number of observed variables being analyzed. In PCA, observed variables are standardized, e.g., mean=0, standard deviation=1, diagonals of the matrix are equal to 1. The amount of variance explained is equal to the trace of the matrix (sum of the diagonals of the decomposed correlation matrix).
6. The number of principal components is less than or equal to the number of original variables. This transformation is defined in such a way that the first principal component has the largest possible variance (that is, accounts for as much of the variability in the data as possible), and each succeeding component in turn has the highest variance possible under the constraint that it is orthogonal to (*i.e.,* uncorrelated with) the preceding components. The principal components are orthogonal because they are the eigenvectors of the covariance matrix, which is symmetric.
7. PCA is sensitive to the relative scaling of the original variables.

8. The number of components extracted is equal to the number of observed variables in the analysis. The first principal component identified accounts for most of the variance in the data. The second component identified accounts for the second largest amount of variance in the data and is uncorrelated with the first principal component and so on.
9. Components accounting for maximal variance are retained while other components accounting for a trivial amount of variance are not retained. Eigen values indicate the amount of variance explained by each component. Eigenvectors are the weights used to calculate components scores.
10. The results of a PCA are usually discussed in terms of component scores, sometimes called factor scores (the transformed variable values corresponding to a particular data point), and loadings (the weight by which each standardized original variable should be multiplied to get the component score).

THE OBJECTIVE OF PRINCIPAL COMPONENT ANALYSIS

1. To reduce the dimensionality (number of variables) of the datas et but retain as much of the original variability in the data as possible. The first principal component accounts for the majority of the variability in the data, the second principal component accounts for the majority of the remaining variability, and so on.
2. It measures divergence between varieties in terms of spatial distance rather than quantifying it as in D^2 analysis. It gives group constellations as varietal distribution in two–way pictorial graph fixes the relative position of each variety.

It also indicates the characters (Variables) most important for divergence at each axis. The largest element in each vector constitutes the greatest contributors.

Analysis of PCA is to be carried out according to the procedure given by Nandarajan and Gunasekaran.

18 Factor Analysis

Factor analysis is a variable reduction technique which identifies the number of latent constructs and the underlying factor structure of a set of variables. It hypothesizes an underlying construct, a variable not measured directly and estimates factors which influence responses on observed variables. In FA, observed variables are a linear combination of the underlying factors (estimated factor and a unique factor). Communality is the variance of observed variables accounted for by a common factor. Large communality is strongly influenced by an underlying construct

SIMILARITIES OF PCA AND FA

PCA and EFA have these assumptions in common:

1. Measurement scale is interval or ratio level
2. Random sample – at least 5 observations per observed variable and at least 100 observations.
3. Larger sample sizes recommended for more stable estimates, 10–20 observations per observed variable
4. Over sample to compensate for missing values
5. Linear relationship between observed variables
6. Normal distribution for each observed variable
7. Each pair of observed variables has a bivariate normal distribution
8. PCA and FA are both variable reduction techniques. If communalities are large, close to 1.00, results could be similar

Differences

Principal Component Analysis	Factor Analysis
Principal Components retained account for a maximal amount of variance of observed variables	Factors account for common variance in the data
Analysis decomposes correlation matrix	Analysis decomposes adjusted correlation matrix
Ones on the diagonals of the correlation matrix	Diagonals of correlation matrix adjusted with unique factors

Contd...

Contd...

Minimizes sum of squared perpendicular distance to the component axis	Estimates factors which influence responses on observed variables
Component scores are a linear combination of the observed variables weighted by eigen vectors	Observed variables are linear combinations of the underlying and unique factors

19 Combining Ability

In early days Breeders used to select the parents for hybridization on the basis of their *per se* performance or sometimes as arbitrarily. Consequently some crosses appeared to be good and others did not. The knowledge of gene action is essential to a plant breeder for deciding an efficient breeding procedure for the genetic improvement of the characters. The combining ability is a measure of gene action. This was proposed by Sprague and Tatum (1942) for their study on single crosses of Maize.

Combining ability: Combining ability is the capacity or a ability of genotype to transmit its superior or desirable performance to its crosses or progenies.

MAIN FEATURES OF COMBINING ABILITY

1. Combining ability analysis helps in the evaluation of inbreds in terms of their genetic value.
2. Helps in the selection of parents on the basis of gca effects for hybridization and also superior cross combinations on the basis of sca effects which may be utilized for development of superior varieties in self pollinated crops and hybrids in cross pollinated crops.
3. It is useful tool in the development of the synthetic varieties in cross pollinated crops.
4. Combining ability analysis provides information about nature and magnitude of various types of gene action involved in the expression of various quantitative characters and thus it helps in deciding the breeding procedures for genetic improvement of such traits.

TYPES OF COMBINING ABILITY

There are two types of combining ability *viz.*, General combining ability (GCA) and specific combining ability (SCA)

(*i*) **General combining ability (GCA):** The average performance of strain or genotype in a series of crosses is termed as general combining ability. It is estimated from half sib family. It helps in the selection of suitable parents for hybridization. The gca variance is primary a function of the additive genetic variance, but if epistasis is present gca also include additive × additive type of non–allelic interaction. It has relationship with narrow sense heritability. It is important in previously unselected material.

(*ii*) **Specific Combining Ability (SCA):** It is the performance of a specific cross and is often expressed as a deviation of performance predicted by general combining ability. It is estimated from full–sib family. It helps in the selection of superior cross combinations. The sca variance is primary a function of the dominance genetic variance, but if epistasis is present sca also include additive × dominance and dominance × dominance non– allelic interactions. It has relationship with heterosis and is more important in selected material.

MATING DESIGNS USED FOR ESTIMATION OF COMBINING ABILITY

Mating design: It is a system of mating employed to develop progenies of certain kinds. In plant breeding, various mating designs and arrangements are used by breeders and geneticists to generate improved plants. The selection of suitable parents and good mating designs are keys to the successful plant breeding schemes. However, there are several factors affecting the choices of mating designs. The factors influencing the choice of mating design are the type of pollination (self– or cross–pollinated), the type of crossing to be used (artificial or natural), the type of pollen dissemination (wind or insect), the presence of a male–sterility system, the purpose of the project (for breeding or genetic studies) and the size of the population required

The choice of a mating design for estimating genetic variances should be dictated by the objectives of the study, time, space, cost and other biological limitations.

Different methods of mating designs have been developed to estimate the general and specific combining abilities by various workers. Cockerham (1963) classified designs as one, two, three and four factor designs depending upon number of ancestors per progeny.

(*i*) One factor:
1. Top cross (Davis, 1927),
2. Poly cross (Tysdal *et al.,* 1942)

(*ii*) Two factor:
1. Diallel mating design (Griffing,1956)
2. Line × Tester mating design (Kempthorne, 1957).
3. Partial diallel design (Kempthorne and Curnow, 1961)
4. Random mating designs

(*i*) Bi–parental mating *i.e.* North Carolina designs– I, II and III (Comstock and Robinson, 1948 and 1952)

(*ii*) Triple test cross (Kearsy and Jinks, 1968)

Three factor : Triallel analysis (Rawlings and Cockerham, 1962a)

Four factor: Quadriallel analysis (Rawlings and Cockerham, 1962b)

20

Top Cross or Inbred–variety Cross

Top cross or inbred –variety cross refers to a mating between a selection, line, clone and a common pollen parent which may be a variety preferably open pollinated with broad genetic base. Davis (1927) was the first to suggest the use of top cross as method of evaluating inbred lines. The selected plants are crossed with a common tester of known performance, generally in open pollination. The purpose of using top cross is to increase the chance of obtaining a desirable gene or genes from exotic or difficult materials.

Top cross has been fairly used for preliminary evaluation of combining ability of new inbred lines. The possible numbers of crosses are $n \times 1$, given n number of inbreds. Top cross progenies yield only GCA information, not SCA. It is a simple and efficient system of screening inbred lines for combining ability before pairing inbreds in single–cross yield trials. This design is probably the simplest model of mating design that can provide preliminary rapid screening of genetic stocks as it involves the lowest crossing load and simple statistical analysis. In top cross, the progenies from individual plants are tested; these progenies are half–sib families. The covariance within the families is $Cov\ (HS) = \frac{1+F}{4}\sigma^2_A$ where F the inbreeding coefficient of the genotypes is tested.

EFFECTING CROSSES, EVALUATION AND STATISTICAL ANALYSIS

Select number of test indreds (n) as female parents and crosses are made using single wide based tester variety as a pollen parent. These crosses along with inbreds are evaluated in RBD with required number of replications.

The mean data recorded on quantitative characters on all genotypes are subjected to the simple analysis of variance for top crosses and inbred parent. And ANOVA is as below.

Analysis of variance for Top crosses

Sources	Df	SS	MS
Replications	$(r-1)$		
Genotypes	$(g-1)$		
Parents	$(p-1)$		
Top crosses	$(c-1)$		
P Vs C	1		
Error	$(r-1)(g-1)$		
Total	$(rg-1)$		

21 Polycross Design

This design is for inter mating a group of cultivars by natural crossing in isolated block. Term polycross was coined by Tysdal, Kiesselbach and Westover (1942), to indicate progeny from seed of a line that was subject to out crossing with other selected lines growing in the same nursery. It is most suited to species that are obligate cross–pollinators (e.g., forage grasses and legumes, sugarcane, sweet potato), but especially to those that can be vegetatively propagated crops such as sugarcane, cassava and sweet potato. The design provides equal opportunity for each and every clone or parent to naturally cross with each other in the block such that self–pollination is prevented.

The Latin square experimental design is the most appropriate design to ensure all entries have equal chance of random intermating with each other in the polycross nursery. Nevertheless, when the entries number is more than 10, the completely randomized block design may be used. In both cases, about 20–30 replications are included in the crossing block.

In polycross, the progenies from individual plants are tested. These progenies are half–sib families. The mean performance of the progenies of any female parent in the polycross is used to determine the variance components and consequently the general combining ability (gca). The variance component σ^2 progenies is an estimate of *Cov (HS)* $= \frac{1+F}{4}\sigma_A^2$ when the parents are non–inbred, *F*= zero.

The design is used in breeding to produce synthetic cultivars, recombining selected entries of families in recurrent selection breeding programs, or for evaluating the GCA of entries.

EFFECTING CROSSES, EVALUATION AND STATISTICAL ANALYSIS

After random mating and at maturity, seed of each of test genotypes should be collected from the polycross nursery, evaluate all the polycross progenies along with parental seeds of each inbred in replicated RBD. Data are collected and statistical analysis is done in the same fashion as in the case of top crosses.

However, polycross design has a number of limitation such as random mating, insufficient statistics to estimate all the parameters, the component of variance are only estimated from the maternal half sibs; information about the males is lost, no control over the pollen source; expected genetic gains are reduced by half, the non–randomness of mating (due to lack of synchronization of flowering, unequal pollen production and position effects in the crossing block).

22

Line × Tester Analysis

The concept of line × tester analysis was developed by Kempthorne (1957). It is a extension or modified form of top cross scheme. In case of top cross only one tester is used, while in line × tester cross several testers used.

Selection of parents and effecting crosses

In this design, the genotypes for crossing are selected from the germplasm. Some of the selected genotypes usually inbreds, newly developed cultures, varieties are designated as 'lines' (females). Some of selected genotypes, usually the wider adapted varieties, inbreds and varieties having broad genetic base are designated as 'tester' (males). Usually lines are more in number and testers are to be less in number (but not less than three).

Each male is crossed to each female parent, but male parent is not crossed to each other and female parents are not crossed to each other. Hence, the total number of crosses are = males × females.

MAIN FEATURES OF LINE × TESTER TECHNIQUE

1. L × T technique is applicable for both self and cross pollinated species.
2. Both full sibs and half sibs are formed.
3. This technique can evaluate large number of germplasm lines and is useful in deciding the relative ability of a number of female and male inbreds to produce desirable hybrid combination.
4. The genetic information on GCA and SCA variances, effects and additive (D) and dominance (H) components can be obtained.
5. Generally it is used where female inbred having male sterility.

Evaluation of Crosses

The crosses obtained by L × T mating are evaluated in standard design (normally RBD) with required number of replications. The biometrical observations are taken over all the replications and are used for statistical analysis.

Modified line × tester

Usually only crosses are evaluated in L × T proposed by Kempthorne (1957). However, breeders are interested to study the heterosis of crosses. Therefore, parents (lines as well as testers) and some times standard check (hybrid/variety) are also included along with the crosses and all are evaluated in a same experiment. Significance of P vs C is taken as indication of presence of heterosis. Although this method is statistically valid, but conclusion regarding heterosis may not be reliable and useful. It is comparison of mean of parents a group with the mean of hybrids a group. Significance could result even when mean of hybrids as a group is smaller than mean of parental lines as a group and thus can mislead the interpretations. More ever, even P vs C are found significant, many crosses do not show desirable heterosis. Further, the F test with 1 df is known to be less efficient than those with large df. When hybrids (mf) are raised along with parents ($m + f$), Arunachalam (1974) called it as modified Line × Tester in which (m + F +MF) genotypes are raised in required replications. In this modified L × T design, Crosses alone should be randomized in the blocks in an experiment and parents could be raised in an adjacent block. The data on crosses could be used for combining ability analysis and data on parents could use for estimation of heterosis.

Line × Testers Design	Modified L × T Design
Comprises mf crosses only	Comprises Parents + mf crosses
There is no provision of inclusion of parents.	Parents and hybrids are evaluated simultaneously so heterosis can be estimated
Error $df = (r-1)(lt-1)$	Error $df = (r-1)(lt+l+t-1)$
Parents Vs. crosses variation is not possible	Parents Vs. crosses variation is possible
gca effect of female parent is (gi) GCA effect of male parent is (gj)	gca effect of female is $fi + gi$ gca effect of male is $mi + gj$

STATISTICAL ANALYSIS

(*i*) **Analysis of Variance:**

The mean data of quantitative characters recorded on all the genotypes are subjected to appropriate analysis of variance of RBD. The significance of F value for genotypes indicate significant difference among the genotype studied and statistical analysis may be continued for combining ability analysis.

(*ii*) Analysis of variance for combining ability

The degrees of freedom for various sources of variation for combining ability analysis with parents and without parents in the experiment are as below

Without parents		With parents Modified L × T	
Source	***df***	**Source**	***df***
Replication	$(r-1)$	Replication	$(r-1)$
Lines	$(l-1)$	Genotypes	$(g-1)$
Testers	$(t-1)$	Parents	$(p-1)$
Lines × Testers	$(l-1)(t-1)$	Line	$(l-1)$

Contd...

Contd...

Error	$(r-1)(lt-1)$	Tester	$(t-1)$
Total	$(rlt-1)$	Line vs tester	1
		Crosses	$(lt-1)$
		P vs C	1
		Lines	$(l-1)$
		Testers	$(t-1)$
		Lines × Testers	$(l-1)(t-1)$
		Error	$(r-1)(lt+l+t-1)$
		Total	$[r(lt+1+t)]-1$

The ANOVA for combining ability

Source	*df*	SS	MS	Expectation of Mean squares
Crosses	$(lt-1)$	CSS		
Lines	$(l-1)$	LSS	MS_1	$\sigma^2_e + r$ (Cov. FS − 2 Cov. HS) + rt(Cov. HS)
Testers	$(t-1)$	TSS	MS_2	$\sigma^2_e + r$ (Cov. FS − 2 Cov. HS) + rl(Cov. HS)
Lines × Testers	$(l-1)(t-1)$	Lx TSS	MS_3	$\sigma^2_e + r$ (Cov. FS − 2 Cov. HS)
Error	$(r-1)(lt-1)$	ESS	MS_4	σ^2_e
Total	$(rlt-1)$			

Genetic Information

The general and specific combining ability variances and effects can be estimated by this technique

(*a*) Combining Ability Variances:

Variance of GCA = Cov. HS (Covariance of Half sibs)

$$= \frac{(MS_1 - MS_3) + (MS_2 - MS_3)}{r\,(l+t)}$$

Variance of SCA = Cov. FS –2 Cov. HS

$$= \frac{MS_1 + MS_2 + MS_3 - 3MS_4}{3\,r} + \frac{6r\text{Cov.HS} - r\,(1+t)\,\text{Cov.HS}}{3r}$$

$$\text{Cov FS} - 2\text{Cov. HS} = \frac{(MS_3 - MS_4)}{r}$$

When the parents are inbreds or pure lines (Inbreeding Coefficient, F=1)

$$\text{Cov. HS} = \frac{1}{2}\sigma^2_D + \frac{1}{4}\sigma^2_{DD} \text{ other forms of additive epistasis}$$

$$\text{Cov. FS} = \sigma^2_D + \sigma^2_H + \sigma^2_{DD} + \sigma^2_{DH} + \sigma^2_{HH} + \text{other forms of epistasis}$$

Assuming there is no epistasis,

σ^2_D (Additive genetic variance) = 2Cov.HS or 2 σ^2GCA

σ^2_H (Dominance genetic variance) = Cov. FS– 2Cov.HS or σ^2SCA

When the parents are non– inbreds or cross pollinated parents (Inbreeding Coefficient, F = 0)

$$\text{Cov. HS} = \frac{1}{4}\sigma^2_D + \frac{1}{16}\sigma^2_{DD} \text{ other forms of additive epistasis}$$

$$\text{Cov. FS} = \frac{1}{2}\sigma^2_D + \frac{1}{4}\sigma^2_H + \frac{1}{4}\sigma^2_{DD} + \frac{1}{8}\sigma^2_{DH} + \frac{1}{16}\sigma^2_{HH} \text{ other forms of epistasis}$$

Assuming there is no epistasis,

σ^2_D (Additive genetic variance) = 4Cov.HS or 4 σ^2GCA

σ^2_H (Dominance genetic variance) = 4[Cov. FS – 2Cov.HS] or 4σ^2SCA

(*b*) Combining Ability Effects:

$$\text{GCA effects of lines } (g_i) = \frac{x_{i...}}{xt} - \frac{x_{i...}}{xlt}$$

$$\text{GCA effects of testers } (g_j) = \frac{x_{.j.}}{rl} - \frac{x_{i...}}{rlt}$$

$$\text{SCA effects of hybrids } (S_{ij}) = \frac{x_{ij}}{r} - \frac{x_i}{rt} - \frac{x_{.j.}}{rl} - \frac{x_{i...}}{rlt}$$

Where,

$x\ldots$ – Total of all hybrids over replications

$x_{i.}$ – Total of ith line over t testers and r replications

$x_{.j..}$ – Total of jth tester over l lines and r replications

$x_{ij..}$ – Total of of the hybrid between i^{th} line and jth tester over r replications

Significance of effects is tested by 't' test

$$\text{SE for gca of lines} = \sqrt{\frac{EMS}{rt}}$$

$$\text{SE for gca of testers} = \sqrt{\frac{EMS}{rl}}$$

$$\text{SE for sca of hybrids} = \sqrt{\frac{EMS}{r}}$$

$t = \frac{\text{Effect}}{\text{SE}}$, Calculated 't' value can be compared with table 't' value at error df

$$\text{SE } (g_i - g_j) \text{ lines} = \sqrt{\frac{2EMS}{rt}}$$

$$\text{SE } (g_i - g_j) \text{ testers} = \sqrt{\frac{2EMS}{rl}}$$

$$\text{SE } (s_{ij} - s_{kl}) \text{ hybrids} = \sqrt{\frac{2EMS}{r}}$$

CD = SE × t value at error df at 5 % and 1 % level of significance.

(c) Percent contribution of lines, testers and their interaction to total variance

$$\text{Contribution of lines} = \frac{SS(\text{lines})}{SS(\text{crosses})} \times 100$$

$$\text{Contribution of testers} = \frac{SS(\text{testers})}{SS(\text{crosses})} \times 100$$

$$\text{Contribution of } (l \times t) = \frac{SS(l \times t)}{SS(\text{crosses})} \times 100$$

SOLVED EXERCISE

A total of 12 sorghum hybrids obtained by L × T mating design (4 lines and 3 testers) and 7 parents were evaluated in a RBD with three replications. The data on grain yield is given below. Estimate gca and sca effects and workout the mode of inheritance and suggest suitable breeding method to improve grain yield.

Sr.		Seed yield/plant (g)				
No.	Crosses	R–I	R–II	R–III	Total	Mean
1	L1 × T1	100.40	90.90	88.00	279.30	93.10
2	L1 × T2	92.80	99.20	102.30	294.30	98.10
3	L1 × T3	98.10	82.10	86.20	266.40	88.80
4	L2 × T1	78.20	80.30	75.90	234.40	78.13
5	L2 × T2	88.10	90.20	85.40	263.70	87.90
6	L2 × T3	87.00	79.00	82.30	248.30	82.77
7	L3 × T1	103.10	109.70	119.40	332.20	110.73
8	L3 × T2	95.60	100.90	99.10	295.60	98.53
9	L3 × T3	93.60	98.30	84.60	276.50	92.17
10	L4 × T1	78.60	73.10	75.90	227.60	75.87
11	L4 × T2	81.40	86.80	83.70	251.90	83.97
12	L4 × T3	86.60	94.20	85.40	266.20	88.73
	Crosses Total				3236.4	
13	L1	94.90	84.30	86.40	265.60	88.53
14	L2	77.90	71.30	69.50	218.70	72.90
15	L3	90.80	96.50	95.90	283.20	94.40
16	L4	60.00	52.50	51.00	163.50	54.50
17	T1	77.40	78.80	80.20	236.40	78.80
18	T2	91.40	85.00	89.50	265.90	88.63
19	T3	91.80	84.80	81.50	258.10	86.03
	Total	1667.70	1637.90	1622.20	4927.80	

1. **Analysis of variance.**

$$\text{Correction Factor} = \frac{(4927.80)_2}{57} = 426021.28$$

$$\text{TSS} = 434406.86 - \text{CF} = 8385.58$$

$$\text{Treat. SS} = 433513.69 - \text{CF} = 7492.41$$

$$\text{Repl.SS} = 426077.50 - \text{CF} = 56.22$$

$$\text{Error SS} = \text{TSS} - \text{Treat.SS} - \text{Repl.SS} = 836.95$$

ANOVA for RBD

Sources	Df	SS	MS	F
Replications	2	56.22	28.11	1.21
Treatments	18	7492.41	416.24**	17.90
Error	36	836.95	23.25	

2. **Partitioning of treatment SS**

For this purpose, the grand total for crosses, parents, lines, testers determined separately

$$\text{SS of Crosses} = \frac{\Sigma\Sigma C_{ij}^{2}}{r} - \text{CF (crosses) with 11 } d.f.$$

$$\text{CF Crosses} = \frac{(\text{Crosses total})^2}{\text{No. of crosses} \times \text{No. of relp.}} = \frac{(3236.4)^2}{12 \times 3}$$

$$= \frac{10474284.96}{36} = 290952.36$$

$$\text{SS Crosses} = \frac{(882030.14)}{3} - 290952.36 = 3057.68$$

$$\text{SS of Parents} = \frac{\Sigma\Sigma P_{i}^{2}j}{r} - \text{CF (Parents) with 6 } d.f.$$

$$\text{CF Parents} = \frac{(2860833.96)}{21} = 136230.18$$

$$\text{SS Parents} = \frac{(418510.92)}{3} - 136230.18 = 3273.45$$

$$\text{SS of lines} = \frac{\Sigma\Sigma I_{i}^{2}j}{r} - \text{CF (lines) with 3 } d.f.$$

$$\text{CF Lines} = \frac{(866761.0)}{12} = 72230.08$$

$$\text{SS Lines} = \frac{(225307.54)}{3} - 72230.08 = 2872.43$$

$$\text{SS of Tester} = \frac{\Sigma\Sigma r_{i}^{2}j}{r} - \text{CF (Parents) with 2 } d.f.$$

$$\text{CF Testers} = \frac{(578208.16)}{9} = 64245.35$$

$$\text{SS Testers} = \frac{(193203.38)}{3} - 64245.35 = 155.77$$

SS (Lines vs. Testers) = SS Parent – Line SS Line – SS Testers with 6 – 3 – 2 = 1 *d.f.*

= 3273.45 – 2872.43 – 155.77 = 245.24

SS (Parent vs. crosses) = Treat.SS – Crosses SS – Parent SS with 18 – 11 – 6 = 1 *d.f.*

= 7492.40 – 3057.68 – 3273.45 = 1161.27

3. Line × Tester analysis

Prepare a two way table of data on crosses as below

Two way Table

Testers/Lines	T_1	T_2	T_3	x_i
L1	279.30	294.30	266.40	**840.00**
L2	234.40	263.70	248.30	**746.40**
L3	332.20	295.60	276.50	**904.30**
L4	227.60	251.90	266.20	**745.70**
$x_{.j.}$	**1073.50**	**1105.50**	**1057.40**	**3236.40**

Work out following sum of squares

$$\text{SS due to lines} = (840.00)^2 + ... + \frac{(745.70)^2}{3} \times 3 - \text{CF of Crosses}$$

$$= 292948.88 - 290952.36 = 1996.52$$

$$\text{SS due to Testers} = (1073.50)^2 + ... + \frac{(1057.40)^2}{3} \times 4 - \text{CF of Crosses}$$

$$= 291052.27 - 290952.36 = 99.91$$

$$\text{SS due to } (L \times T) = \text{SS crosses} - \text{SS Lines} - \text{SS Testers}$$

$$= 3057.68 - 1996.52 - 99.91 = 961.25$$

ANOVA for L × T analysis

Sources	df	SS	MS		F cal
Lines	3	1996.52	665.51**	(MS_1)	4.15
Testers	2	99.91	49.96	(MS_2)	0.31
Line × Tester	6	961.25	160.21**	(MS_3)	5.59
Error	22	629.96	28.63	(MS_4)	

ANOVA for L × T analysis (Modified L × T)

Sources	df	SS	MS	F cal
Replications	2	56.22	28.11	1.21
Treatments	18	7492.41	416.24**	17.90
Parents	6	3273.4514	545.58**	23.47
Lines	3	2872.43	957.48**	41.18
Testers	2	155.77556	77.89*	3.35
Line vs. Tester	1	245.24587	245.25**	10.55
Crosses	11	3057.6867	277.97**	11.96
Parent Vs. Crosses	1	1161.2707	1161.27**	49.95
Line effect	3	1996.52	665.51**	4.15
Tester effect	2	99.91	49.96	0.31
Line × Tester effect	6	961.25	160.21**	5.59
Error	36	836.95	23.25	

Here, MS due to lines effects and tester effects are tested against MS due to Line × Tester effect. The remaining are tested against MS due to error. From the above ANOVA , it is observed that all the mean squares are found to significant except MS due to tester effect, which indicates that except tester, there are sufficient variability among crosses and lines.

(c) Combining Ability Variances:

$$\text{Variance of GCA } (\sigma^2_{GCA}) = \frac{[(MS_1 - MS_3) + (MS_2 - MS_3)]}{r}(l+t)$$

$$= \frac{[(665.51 - 160.21) + (49.96 - 160.21)]}{3}(4x+3)$$

$$= 34.05$$

$$\text{Additive genetic variance } (\sigma^2_D) = 2 \times \sigma^2_{GCA} = 2 \times 34.05 = 68.10$$

$$\text{Variance of SCA } (\sigma^2_{SCA}) = \frac{[(MS_3 - MS_4)}{r} = \frac{(160.21 - 28.63)}{3} = 43.86$$

$$\text{Dominance genetic variance } ((\sigma^2_H) = \sigma^2_{SCA} = 43.86$$

From the combining ability variances, it is noticed that additive genetic variance (σ^2_D) is greater than dominance genetic variance (σ^2_H) which indicates the trait grain yield in sorghum is governed by additive gene action. Therefore recombination breeding followed by pedigree selection is suggested to improve grain yield.

ESTIMATION OF COMBINING ABILITY EFFECTS

(i) General combining ability effects (GCA)

$$\text{gca effects of lines } (g_i) = \frac{x_{i...}}{rt} - \frac{x_{i...}}{xlt}$$

$$\text{SE for gca of lines} = \sqrt{\frac{EMS}{rt}} = \sqrt{\frac{23.25}{3 \times 3}}\ 1.606$$

$$\text{gca effects of testers } (g_j) = \frac{x_{j...}}{rl} = \frac{x_{i...}}{rlt}$$

$$\text{SE for gca of testers} = \sqrt{\frac{EMS}{rl}} = \sqrt{\frac{23.25}{3 \times 4}} = 1.391$$

Parents	GCA effect	$t = \frac{\text{Effect}}{\text{SE}}$
Lines		
L_1	$\frac{840.0}{3 \times 3} - \frac{3236.40}{3 \times 4 \times 3} = 3.43^*$	$\frac{3.43}{1.606} = 2.14$
L_2	–6.97**	–4.33
L_3	10.58**	6.58
L_4	–7.04**	–4.38
Testers		
T_1	$\frac{1073.15}{3 \times 4} - \frac{3236.40}{3 \times 4 \times 3} = -0.44$	$\frac{-0.44}{1.392} = -0.32$
T_2	2.22	1.59842
T_3	–1.78	–1.2811

$$\text{sca effects of hybrids } (S_{ij}) = \frac{x_{ij}}{r} - \frac{x_i}{rt} - \frac{x_j}{rl} - \frac{x_i}{rlt}$$

$$\text{SE for sca of hybrids} = \sqrt{\frac{EMS}{r}} = \sqrt{\frac{23.25}{3}} = 2.784$$

Hybrids	SCA effect		$t = \frac{\text{Effect}}{\text{SE}}$
$L_1 \times T_1$	$\frac{279.30}{3} - \frac{840.0}{3 \times 3} - \frac{1073.50}{3 \times 4} + \frac{3236.40}{3 \times 4 \times 3}$	0.21	$\frac{0.21}{2.784} = 0.07$
$L_1 \times T_2$		2.54	0.91
$L_1 \times T_3$		–2.75	–0.98
$L_2 \times T_1$		–4.36	–1.56
$L_2 \times T_2$		2.74	0.98
$L_2 \times T_3$		1.62	0.58
$L_3 \times T_1$		10.69**	3.84
$L_3 \times T_2$		–4.18	–1.50
$L_3 \times T_3$		–6.53*	–2.34

Contd...

Contd...

$L_4 \times T_1$	–6.55*	
$L_4 \times T_2$	–1.12	
$L_4 \times T_3$	7.65**	

The '*t*' calculated value is greater than *t* table value at error *df* (36) *i.e.* 2.03 and 2.72 at 5 and 1 % level, the estimate is considered as significant.

Proportional contribution of lines, testers and their interaction to total variance

$$\text{Contribution of lines} = \frac{\text{SS(lines)}}{\text{SS(crosses)}} \times 100 = \frac{1996.52}{3057.68} \times 100 = 65.30$$

$$\text{Contribution of testers} = \frac{\text{SS(testers)}}{\text{SS(crosses)}} \times 100 = \frac{99.91}{3057.68} \times 100 = 3.26$$

$$\text{Contribution of } (l \times t) = \frac{\text{SS}(l \times t)}{\text{SS(crosses)}} \times 100 = \frac{961.25}{3057.68} \times 100 = 31.44$$

CONCLUSION

The analysis of variance for combining ability revealed that MSS due to lines, crosses, L × T were highly significant for grain yield whereas non significant for testers. This indicated that there was substantial variability among the lines chosen for crossing programme.

Additive variance was higher in magnitude than dominance variance indicating predominant role of additive genetic system in the inheritance of grain yield.

The estimates of gca effects revealed that the line L_1 and L_2 had positive and significant gca effects. Therefore these lines are good combiners. None of the testers had significant gca effect. The hybrids $L_3 \times T_1$ and $L_4 \times T_3$, recorded significant positive sca therefore these hybrids are recommended for heterosis breeding after testing over seasons and years.

23 Diallel Analysis (Hayman's Graphical Approach)

Diallel cross refers to mating of selected parents in all possible combinations. A evaluation of a set of diallel crosses is known as diallel analysis. The diallel mating design is used to evaluate several inbred lines in terms of combining ability variances and effects and also provides information on nature and amount of genetic parameters. There are two approaches of diallel analysis *viz.*, Hayman' s graphical approach and Griffing's numerical approach.

HAYMAN'S GRAPHICAL APPROACH

The theory and analysis of graphical approach of diallel has been explained initially by Jinks and Hayman (1953) and later on elaborated independently by Jinks (1954 and 1956) and Hayman (1954 a, b).

Hayman (1954) proposed additive–dominance model for diallel genetic analysis. In this model only additive and dominance genetic parameters are estimated. To fulfill the above genetic model the following are the assumptions.

1. **The inheritance diploid segregation:** This assumption is usually fulfilled as most of the times we are dealing diploid population or allopolyploids that behave like diploids during meiosis. Diallel analysis of auto polyploids should be avoided.
2. **Absence of multiple alleles:** Multiple alleles arise due to mutations, but mutation occurs at very low frequency, thus multiple alleles is not at problem.
3. **Absence of maternal effects:** If maternal effects are present, analysis needs some modification.
4. **Absence of G × E interaction:** This assumption is nit fulfilled by many times, because it is not in control of plant breeder. Growing of material in large number of environments is necessary to minimize the G × E interaction but mostly this is not followed.
5. **The parents are completely homozygous or inbreds:** Parents selected should be homozygous.
6. **Independent distribution of genes between the parents (No linkage):** In diallel analysis it is assumed that genes are independently distributed among the parents. Failure of this assumption results from presence of linkage.

7. **Independent action of non allelic gene interactions (Absence of epistasis):** In most of the cases there is correlation between heterosis and presence of epistasis, we can not rule out the presence of epistasis.

Steps of statistical analysis

1. **Testing of Model:**

To test the adequacy of additive–dominance model there are two important tests as given below

(*i*) t^2 test or F test:

(*ii*) Regression coefficient test

***(i)* t^2 test or F test:** The uniformity of W_r, V_r indicates the validity of assumptions made by Hayman (1954).

$$t^2 = \frac{(n-2)}{4}\left[\frac{(VarV_r - VarW_r)^2}{VarV_r \times VarW_r - Cov(V_r - W_r)}\right]$$

Where, n = No. of parents

Test with F value for n and $(n-2)$ degree of freedom. If the t^2 is significant, the additive dominance model is not good fit. Then it is necessary to extend the model with additional generation other than parents and F_1.

***(ii)* Regression coefficient test:** If there is interactions (if the contributions of genes are not independent), we can no longer except the regression of *Wr* on *Vr* to be rectilinear with the slope of 1

$$b = \left[\frac{Cov(V_r, W_r)}{Var\ V_r}\right]$$

The b value should not significantly deviate from 1

$$\text{SE}b = \sqrt{\frac{\frac{EMS}{SSVr}}{(n-2)}}$$

Calculate $$t = \frac{b-1}{\text{SEB}}$$

Different parameters to be worked out from Diallel table

V_r	Variance of progeny family means within an array (r[th] array)
$V_{\bar{r}} = V_{1L1}$	Mean of variance of an array (r[th] array)
$V_p = V_{0L0}$	Variance of parental array (r[th] array)
$\overline{V}_r = V_{0L1}$	Variance of mean of array
W_r	Covariance between parents and their offspring in one array (r[th] array)

Contd...

Contd...

$\overline{W}r = W_{0L1}$	Mean of covariance between parents and their offspring in one array (r^{th} array)
$E = M_e'$	The expected environmental component of variation for parents and crosses (F_1s) data of individual environment
$(m_{L1} - m_{L0})^2$	The difference between mean of the parents and mean of their n^2 progenies

3. Estimation of Genetic Components

(*i*) E: Environmental Component E = EMS/r

where EMS is error mean square in ANOVA and *r* is number of replications

(*ii*) D: Additive Genetic Variance $\hat{D} = V_p - \hat{E}$

(*iii*) H_1: Dominance variance component of variation due to dominant gene effect

(*iv*) $\hat{H}_1 = 4\bar{V}_r + V_p - 4\bar{W}_r - \left(\frac{3n-2}{n}\right)\hat{E}$

H_2 : Symmetry or Asymmetry of alleles $\hat{H}_2 = 4\bar{V}_r - 4V_{\bar{r}} - 2\hat{E}$

(*v*) F: The mean of F*r* over arrays, F*r* being the covariance of additive and non additive

effects in the rth array $\hat{F} = 2V_p - 4\bar{W}_r - \frac{2(n-2)}{n}\hat{E}$

(*vi*) h^2: Dominance effect. It is algebraic sum over all loci in heterozygous phase in all crosses $\hat{h}^2 = 4(m_{L1} - m_{L0})^2 - \frac{4(n-1)}{n^2}\hat{E}$

where,

m_{L1} is the grand mean of all parents and hybrids

m_{Lo} is mean of all parents

n is no. of parents.

Test of significance of each component: SE for each component is calculated as the square root of the products of common multiplier and specific multiplier.

For this common multiplier (S^2) is computed as below

$$S^2 = \frac{1}{2}\left[\frac{1}{n-1}\right]\left[\sum(W_{r1} - V_{r1})^2 - \frac{\sum(W_{r1} - V_{r1})^2}{n}\right]$$

The specific multipliers for each component (n parents) is calculated as below

$$D = \frac{(n_5 + n_4)}{n^5}$$

$$H_1 = \frac{(n^5 + 41n^2 - 12n^3 + 4n^2)}{n^5}$$

$$H_2 = \frac{36n^4}{n^5}$$

$$F = \frac{(4n^5 + 20n^4 - 16n^3 + 16n^2)}{n^5}$$

$$h^2 = \frac{(16n^4 + 16n^2 - 32n + 16)}{n^5}$$

$$E = \frac{n^4}{n^5}$$

Then the significance of each component is tested by t test at $n - 2$ *d.f.*

$$'t' = \frac{\textit{Estimate of Component}}{SE}$$

The above six components help in estimation of various genetic ratios as below

4. Estimation of different ratios :

1. **Average degree of dominance. As action of genes**

$$\text{For } F_1 \text{ data } \sqrt{\frac{H_1}{D}} \text{ and for } F_2 \text{ data } \sqrt{\frac{1}{4} \times \frac{H_1}{D}}$$

If the value of this ratio is 1, it indicates complete dominance; it is less than 1, it indicates partial dominance; if the value is 0, no dominance and if it is more than 1, it indicates over dominance.

2. Proportion of +Ve and –Ve alleles in the parents

$$\frac{H_2}{4H_1} = \text{same for } F_1 \text{ and F2}$$

If the value is 0.25, it indicates symmetry of + Ve and –Ve alleles and if it less than 0.25 , it indicated asymmetry of + Ve and – Ve allele

Proportion of dominant and recessive alleles in the parents

$$\frac{\sqrt{4(DH_1)} + F}{\sqrt{4(DH_1)} - F} = \frac{K_D}{K_R} (\text{for } F_1 \text{ data})$$

$$\frac{\sqrt{\frac{1}{4}(4DH_1)} + \frac{1}{2}F}{\sqrt{\frac{1}{4}(4DH_1)} - \frac{1}{2}F} = \frac{K_D}{K_R} \ (\text{for } F_2 \text{ data})$$

It the ratio is more than 1, it indicates more dominant alleles in the parents; if it is less than 1, it indicates more of recessive alleles and if it is equal to 1 it indicates equal proportion of dominant and recessive alleles in the parents.

4. The number of groups of genes which control the character and exhibit dominance

$$= \frac{h^2}{H_2} \quad \text{(same for both } F_1 \text{ and } F_2\text{)}$$

5. Heritability (Broad sense) :

$$= \frac{\frac{1}{2}D + \frac{1}{2}H_1 - \frac{1}{4}H_2 - \frac{1}{2}F}{\frac{1}{2}D + \frac{1}{2}H_1 - \frac{1}{4}H_2 - \frac{1}{2}F + E}$$

Heritability (Narrow sense) :

$$= \frac{\frac{1}{2}D + \frac{1}{2}H_1 - \frac{1}{2}H_2 - \frac{1}{2}F}{\frac{1}{2}D + \frac{1}{2}H_1 - \frac{1}{2}H_2 - \frac{1}{2}F + E}$$

The above components and genetic ratios are estimated from the data of F_1 generation of diallel cross by Hayman (1954 a,b) and for $_{F2}$ generation by Jinks (1956)

5. Construction of $W_r V_r$ graph

The relationship of W*r* with V*r* provides very useful information. Therefore, Wr values are plotted against the corresponding values of V*r*. In order to draw parabola, the values of parabola limits are calculated by using following formula.

$$Wri = (Vr_i \times V_oL_o)^{1/2}$$

For drawing regression line, we require expected value and is calculated as follow

$$W_{rie} = \overline{W}_r - b\overline{V}_r + bV_r$$

Where, $\overline{W}_r = W_oL_1$ *i.e.*, mean covariance between parent and array

$V_r = V_1L_1$ *i.e.*, mean variance of the arrays

These values are plotted on W*r* – V*r* graph and a straight line is drawn through these points. The point of interception of the regression line with W*r* ordinate ('*a*') is obtained as $a = \overline{W}_r - b\overline{V}_r$.

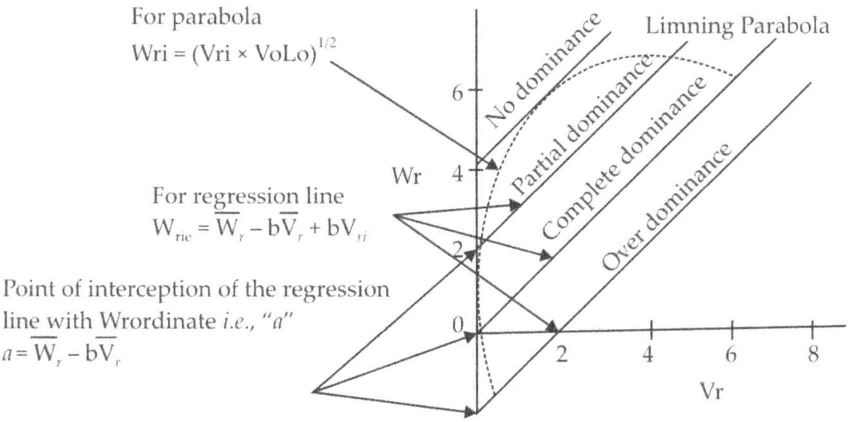

Cited from Singh & Narayanan (2000)

Interpretation from Wr – Vr graph:

1. The position of regression line on graph provides information on level of dominance as
   (*a*) Where D = H_1, the intercept will pass through origin. It indicates complete dominance
   (*b*) Where D > H_1, the intercept will cut the ordinate above the origin. It indicates partial dominance
   (*c*) When the intercept will cut the ordinate above the origin and touches to limiting parabola indicates no dominance.
   (*d*) Where H_1 > D, the intercept will pass below the point of origin. It indicates over dominance
2. The position of array points along regression line indicates
   (*a*) Parents with more of dominant alleles– near the origin on the regression line.
   (*b*) Parents with more of recessive alleles– away from the origin on the regression line.
   (*c*) Parents with dominant and recessive alleles in equal frequencies – Intermediate in position.

SALIENT FEATURES OF HAYMAN'S DIALLEL APPROACH

1. Hayman's analysis is applicable to polygenic systems as well as F_2 families.
2. The component and graphical analysis provides information on additive, non–additive and maternal effects. This approach provides information on selection of breeding methods based on gene action involved
3. It helps in identifying the parents with dominant or recessive gene with positive or negative effects. But this approach does not help in identification of superior combinations
4. When the assumptions are not fulfilled there is scope for exclusion of parents with relative high variance from the analysis.

SOLVED EXERCISE

A trial was conducted with eight parents of maize crossed in a full diallel fashion. The cob length of 64 genotypes (28 straight crosses+28 reciprocals +8 parents) was measured. Mean data on cob length of three replications is analyzed and given below.

Step–I : Analysis of Variance for RBD for cob length (cm)

Sources	df	SS	MSS	F Ratio
Replicates	2	0.804	0.402	0.735
Treatments	63	364.65	5.78**	10.58
Error	126	68.930	0.57	

Significance of F ratio indicated that there was significant differences among the genotypes for cob length and hence proceeded for diallel analysis.

Step–II : ANOVA (according Hayman's) method

Sources	d.f.		MSS
Repl	*r*–1	2	0.402
a = additive genetic effect	P–1	7	25.296**
b = overall dominance effects	P(P–1)/2	28	2.945**
*b*1 = overall direction of dominance	1	1	9.011**
*b*2 = asymmetric distribution of dominant gene in the parents	P–1	7	1.781**
*b*3 = dominance interaction between specific genotypes	P(P–3)/2	20	3.050**
c = overall maternal effects	P–1	7	9.733**
d = residual reciprocal effect	(P–1)(P–2)/2	21	1.760**
Hybrids	$(P^2–1)$	63	5.788**
Error	$(P^2–1)(r–1)$	126	0.547

Step II: Prepare Two –way table of mean over replications

	P1	P2	P3	P4	P5	P6	P7	P8	Total	Mean
P1	13.34	11.69	13.64	14.59	14.17	15.41	12.24	15.05	110.13	18.35
P2		12.44	14.93	14.59	15.13	14.01	12.62	13.19	109.20	18.20
P3			15.20	14.52	15.20	14.36	15.12	14.86	115.51	19.25
P4				14.60	15.55	15.35	14.45	14.18	117.86	19.64
P5					15.34	15.67	13.94	15.95	121.43	20.24
P6						15.70	18.20	15.82	130.68	21.78
P7							14.03	15.74	126.36	21.06
P8								14.54	128.71	21.45
Total									959.88	159.98

Step III: Estimates of variances and co–variances

$$\text{Parental mean} = \frac{1}{8}[13.34 + 12.44 + 15.20 + \ldots. 14.54] = 114.41$$

Variance of parents = $Vp(V_0L_0)$

$$= \frac{1}{p} - 1\frac{[13.34^2 + 12.44^2 + \ldots + 14.54^2 - 115.25]^2}{p} = 1.16$$

Variance of each array – V_r: V_{r1}

$$= \frac{1}{p} - 1\,\frac{[13.34^2 + 11.69^2 + \ldots. + 15.05^2 - 110.13]^2}{8} = 1.73$$

Similarly, $V_{r2} = 1.103$ $V_{r3} = 0.735$ $V_{r4} = 0.210,$ $V_{r5} = 0.0.567$

$V_{r6} = 0.910$ $V_{r7} = 2.283$ $V_{r8} = 0.601$

Variance of the mean of arrays (VoL1)

$$=\frac{1}{p}-1\,\frac{[18.35^2+18.20^2+\ldots.+21.45^2-159.98]^2}{8}$$

Co – variance of each array – W_r: [Covariance between a parent and offspring *i.e.*, covariance of the arrays with non – recurrent parent

$$W_{r1}=\frac{1}{p}-1\,\frac{[(13.34)^2+(11.69\times12.44)+\ldots.+(15.05\times14.54)-(110.13\times114.41)]^2}{8}$$

= 1.208

Similarly, Calculated W_{r2} to W_{r8} and presented in table

Step IV: prepare a table of variance and covariance of arrays

Genotype	Wr	Vr	Wr–Vr	Wr+Vr	Yr	Fr
1 Parent	1.208	1.734	–0.526	2.942	13.34	–3.013
2 Parent	1.038	1.103	–0.065	2.14	12.437	–1.41
3 Parent	0.747	0.735	0.013	1.482	15.197	–0.095
4 Parent	0.255	0.21	0.044	0.465	14.6	1.94
5 Parent	0.46	0.567	–0.107	1.027	15.34	0.816
6 Parent	0.174	0.91	–0.736	1.083	15.937	0.703
7 Parent	1.511	2.283	–0.772	3.794	14.033	–4.718
8 Parent	0.519	0.601	–0.082	1.12	14.543	0.631
Total	5.911	8.143	–2.232	14.054	115.427	–5.145
Mean	0.739	1.018	–0.279	1.757	14.428	–0.643

Step V : Testing the validation of the Hypothesis

The uniformity of W_r, V_r indicates the validity of assumptions made by Hayman (1954). The testing is done by using following way :

$$t^2=\frac{(n-2)}{4}\left[\frac{(\text{VarV}_r-\text{VarW}_r)^2}{\text{VarV}_r\times\text{VarW}_r-\text{Cov}(V_r,W_r)}\right]$$

1.677 *ns* [Estimated Variance of Vr = 0.4631 ;

Variance of Wr = 0.2266; Covariance of Vr, Wr = 0.2866]

The table F value of 4 and $(p-2)$ *df* is 6.16. The non significant of t^2 value indicates the fulfillment of assumption of Hayman.

$$b=\left[\frac{\text{Cov}(V_r,W_r)}{\text{VarV}_r}\right]$$

$$=\frac{0.2866}{0.4631}=0.618\text{ ns}$$

The table t value of 4 and $(p-2)$ *df* is 2.447. The non significant of b value indicates the fulfillment of assumption of Hayman.

Step VI. Estimation of Genetic Components :

Components	Formulae	Estimates
E: Environmental Component	$E = EMS/r$	0.182
D – Additive Genetic Variance	$\hat{D} = V_p - \hat{E}$	1.110
H_1 – Dominance variance component of variation due to dominant gene effect	$\hat{H}_1 = 4\bar{V}_r + V_p - 4\bar{W}_r - \left(\frac{3n-2}{n}\right)\hat{E}$	1.909
H_2 – Symmetry or Asymmetry of alleles	$\hat{H}_2 = 4\bar{V}_r - 4V_{\bar{r}} - 2\hat{E}$	1.600
F – The mean of Fr over arrays, Fr being the covariance of additive and non additive effects in the rth array	$\hat{F} = 2V_p - 4\bar{W}_r - \frac{2(n-2)}{n}\hat{E}$	–0.643
h^2 – Dominance effect. It is algebraic sum over all loci in heterozygous phase in all crosses	$\hat{h}^2 = 4(m_{L1} - m_{L0})^2 - \frac{4(n-1)}{n^2}\hat{E}$	1.234

Steps VII: Estimation of different ratios :

1. Average degree of dominance

$$\sqrt{\frac{H_1}{D}} = \mathbf{1.31}$$

The ratio is more than 1 which indicates that character is governed by over dominance

Proportion of +Ve and –Ve alleles in the parents

$$\frac{H_2}{4H_1} = 0.16$$

The value is less than 0.25 hence it indicates asymmetry of + Ve and –Ve allele

Proportion of dominant and recessive alleles in the parents

$$\frac{\sqrt{4(DH_1)} + F}{\sqrt{4(DH_1)} - F} = \frac{K_D}{K_R} = \mathbf{0.64}$$

The value is less than 1 which indicates more of recessive alleles

4. The number of groups of genes which control the character and exhibit dominance

$$\frac{h^2}{H_2} = 0.77$$

Heritability (Narrow sense) :

$$h^2_{ns} = \frac{\frac{1}{2}D + \frac{1}{2}H_1 - \frac{1}{2}H_2 - \frac{1}{2}F}{\frac{1}{2}D + \frac{1}{2}H_1 - \frac{1}{2}H_2 - \frac{1}{2}F + E} = \mathbf{63.92\%}$$

24

Diallel Analysis (Griffing's Approach)

Griffing (1956) proposed a general procedure for diallel analysis. In this approach mean measurement of a cross is partitioned into two major components (*i*) the contribution of the parents (*GCA*) and*ii*) the excess over and above the sum of the two *gca* effects called (*SCA*) effect.

The advantage of this model is that GCA and SCA variances provide suitable analysis of the predominant role of the additive or non additive gene action. The estimates of *gca* and *sca* effects are useful to locate the parents and crosses that are responsible in bringing about a particular type of gene action. He recognized two models based on the assumptions with regard to sampling nature of experimental materials.

MODEL I (FIXED EFFECT MODEL)

In this model, the experimental material i.e parents are deliberately chosen and the experimental material includes a set of fixed genotypes (inbreds/varieties) as a parents. Inferences are to be drawn on individual line.

MODEL II (RANDOM EFFECT MODEL)

In this model, the experimental material are chosen at random. Hence, it represents the population. Inferences can be made from the sample for the population.

It may be noted that the procedure for procedure for analysis of variance remains the same in both the models, but the expectations of mean squares due to various sources are different. Accordingly, the interpretations of the results also differ.

Type of Diallel Crosses

The diallel mating design involves three types of materials viz., direct crosses or one way crosses, reciprocals and parents. Depending upon number of crosses to be made, diallel cross is of two types.

Full diallel :

In this mating, all the possible crosses are made among the selected parents in both the directions *i.e.* direct and reciprocals. In this, each parent is used as male and female for each mating. Full diallel is used when reciprocal differences are significant and parents do not have male sterility or self incompatibility.

Half diallel :

In this mating, all the possible crosses are made among the selected parents in only one directions *i.e.* direct crosses. In this, each parent is used either as male or as female in the mating. Half diallel is used when reciprocal differences are not significant. It can be used when parents have male sterility or self incompatibility.

Depending on the experimental materials (Parents, direct and reciprocal crosses) involved in the analysis, Griffing has given four methods of diallel analysis as follow:

	Method I	Method II	Method III	Method IV
Material included	Parents, Direct and reciprocal crosses	Parents and Direct crosses	Direct and reciprocal crosses	Direct crosses
No. of crosses	P (P–1)	P (P–1)/2	P (P–1)	P(P–1)/2
No. of genotypes	P^2	P (P+1)/2	P (P–1)	P(P–1)/2
Variances can be calculated	GCA, SCA and RCA variances	GCA and SCA variances	GCA, SCA and RCA variances	GCA and SCA variances
Types of diallel	Full diallel	half diallel	Full diallel without parents	Quarter diallel

STEPS IN DIALLEL MATING DESIGN (GRIFFING'S, 1956)

Selection of Parents: Parents (inbreds, varieties, newly developed cultures etc) are selected based on either model I or Model II.

Effecting Crosses: Selected parents are mated in all possible combinations to obtain crosses. However, crosses are made depending upon method to be used for analysis.

Evaluation of Crosses: The diallel crosses obtained are evaluated in statistical design (RBD) with required number of replications (Depending upon the method involved in the analysis).

Statistical Analysis: The mean data of biometrical characters recorded on the genotypes are subjected to analysis of variance (RBD). The significance of f value for the genotypes indicate significant difference among the genotypes studied and may be continued for combining ability analysis.

Combining Ability ANOVA:

The formulae to work out degrees of freedom to various sources of variation for combining ability analysis with regards to all four methods are as below.

Source					MS	F cal
	Method I	Method II	Method III	Method IV		
GCA	(P – 1)	(P – 1)	(P–1)	(P–1)	M_g	M_g/M_e
SCA	P(P–1)/2	P(P–1)/2	P(P–3)/2	P(P–3)/2	M_s	M_s/M_e
RCA	P(P–1)/2	–	P(P–1)/2	–	M_r	M_r/M_e
Error	(*r*–1) (*t*–1)	(*r*–1) (*t*–1)	(*r*–1) (*t*–1)	(*r*–1) (*t*–1)	M_e	
Total	(P^2–1)	P(P–1)/2–1	P(P–1) –1	P(P–1)/2–1		

The analysis of variance for combining ability along with expectation of mean squares and genetic components for different methods and models are as follow.

ANALYSIS OF VARIANCE FOR COMBINING ABILITY METHOD

The formulae to work out degrees of freedom and sum of squares due to various sources of variation and for combining ability analysis in Method I are as below

Source	*df*	Sum of squares
GCA	P – 1	$\frac{1}{2} P[\Sigma(Yi. + Y.j)^2] - \frac{2}{p^2} Y^2...$
SCA	$\frac{P(P-1)}{2}$	$\frac{1}{2} P[\Sigma\Sigma Yij(Yij + Yji) - \frac{1}{2p}\Sigma(Yi. + Y.j)^2 + \frac{1}{p^2} Y^2...]$
Reciprocal	$\frac{P(P-1)}{2}$	$\frac{1}{2} [\Sigma\Sigma(Yij + Yji)^2]$
Error	$(r-1)(t-1)$	
Total	P^2–1	

Source	MS	Expectation of Mean Squares	
		Model I	Model II
GCA	Mg	$\sigma_e^2 + 2P\left(\frac{1}{p-1}\right)\Sigma g_i^2$	$\sigma_e^2 + \frac{2(P-1)^2}{P}\sigma_s^2 + 2p\sigma_g^2$
SCA	Ms	$\sigma_e^2 + 2P\left(\frac{1}{p(p-1)}\right)\Sigma\Sigma S_{ij}^2$	$\sigma_e^2 + \frac{2(P^2 - P + 1)}{p^2}\sigma_s^2$
RCA	Mr	$\sigma_e^2 + 2P\left(\frac{1}{p(p-1)}\right)\Sigma\Sigma r_{ij}^2$	$\sigma_e^2 + 2\sigma_r^2$
Error	Me′	σ_e^2	σ_e^2

$$Me' = \frac{EMS}{r}$$ he EMS is from analysis of variance of RBD

GENETIC COMPONENTS

Model I

$$\text{Variance due to GCA} = \frac{1}{P-1} = \Sigma g_i^2 = \frac{Mg - Me'}{2P}$$

$$\text{Variance due to SCA} = \frac{2}{P(P-1)} = \Sigma\Sigma S_{ij}^2 = Mg - Me'$$

The ratio of GCA variance/SCA variance indicates the relative significance of additive versus non– additive genetic variance.

Model II

Variance due to GCA = Cov. HS

Variance due to SCA = Cov. FS – 2Cov. HS

$$\text{Variance due to GCA} = \sigma_g^2 = \frac{1}{2P}\left(Mg - \frac{Me' + P(P-1)Ms}{p^2 - P + 1}\right)$$

$$\text{Variance due to SCA} = \sigma_s^2 = \frac{P^2}{2(p^2 - P + 1)}(MS + Me')$$

(*i*) When the parents are inbreds or pure lines (inbreeding Co–efficient, F=1)

$$\text{Cov. HS} = \frac{1}{2}\sigma^2_A + \frac{1}{4}\sigma^2_{AA} + \text{other forms of additive epistasis}$$

$$\text{Cov. FS} = \sigma^2_A + \sigma^2_D + \sigma^2_{AA} + \sigma^2_{AD} + \sigma^2_{DD} + \text{other forms of epistasis}$$

Assuming ther is no epistasis,

σ^2_A (Additive genetic variance) = 2Cov.HS or $2\sigma^2_g$

σ^2_D (Dominance genetic variance) = Cov. FS – 2Cov.HS or σ^2_s

(*ii*) When the parents are non– inbreds (Inbreeding Coefficient, F=0)

$$\text{Cov. HS} = \frac{1}{4}\sigma^2_A + \frac{1}{16}\sigma^2_{AA} + \text{other forms of additive epistasis}$$

$$\text{Cov. FS} = \frac{1}{2}\sigma^2_A + \frac{1}{4}\sigma^2_D + \frac{1}{4}\sigma^2_{AA} + \frac{1}{8}\sigma^2_{AD} + \frac{1}{16}\sigma^2_{DD} + \text{other forms of epistasis}$$

Assuming there is no epistasis,

σ^2_A (Additive genetic variance) = 4Cov.HS or $4\sigma^2_g$

σ^2_D (Dominance genetic variance) = 4[Cov. FS– 2Cov.HS] or $4\sigma^2_s$

The ratio of indicates σ^2_A / σ^2_D the relative significance of additive versus non– additive genetic variance.

Combining Ability Effects

$$\text{gca effects of parents } (g_i) = \frac{1}{2P}(Y_i + Y_j) - \frac{1}{P^2}Y..$$

$$\text{sca effects of hybrids } (Sij) = \frac{1}{2}(Y_{ij} + Y_{ji}) - \frac{1}{2P}(Y_{i\cdot} + Y_{j\cdot} + Y_{i\cdot} + Y_j) \cdot \frac{1}{P^2}Y..$$

$$rca \text{ effects of hybrids } (rij) = \frac{1}{2}(Y_{ij} + Y_{ji})$$

Where,

$Y_{..}$ – Grand total of all hybrids

$Y_{i.}$ – Row total of ith parent

$Y.j_{.}$ – Column total of ith parent

$Y._{i}$ – Column total of jth parent

$Y_{j.}$ – Row total of jth parent

Y_{ij} – Mean value of ij[th] hybrid

Y_{ji} – Reciprocal value of ij[th] hybrid

P – Number of parents involved

Significance of effects is tested by 't' test

$$\text{SE for GCA effect of parents} = \sqrt{\frac{(P-1)}{2P^2} Me'}$$

$$\text{SE for sca effect of hybrids } (sij) = \sqrt{\frac{(P^2 - 2p + 2)}{p^2} Me'}$$

$$\text{SE for rca effect of hybrids } (rij) = \sqrt{\frac{1}{2} Me'}$$

$$SE\ (g_1 - g_j) = \sqrt{\frac{1}{P} Me'}$$

$$SE\ (s_{ij} - s_{ik}) = \sqrt{\frac{(P-1)}{P} Me'}$$

$$SE\ (s_{ij} - s_{kl}) = \sqrt{\frac{(P-2)}{P} Me'}$$

$$SE\ (r_{ij} - r_{kl}) = \sqrt{Me^2}$$

$'t' = \frac{\text{Effect}}{\text{SE}}$ The calculated '*t*' value is compared with '*t*' table at error *df*

SOLVED EXERCISE

Diallel Analysis Griffing (1956) method I

A six inbreds of pearl millet were crossed in diallel fashion following Method I. The data on ear head length was measured in three replication for 36 genotypes of pearl millet (15 straight+15 reciprocal crosses and 6 parents). The analysis is done as follows.

Sr.No	Genotype	Ear head length (cm)				
		RI	RII	RIII	Total	Mean
1	$P_1 \times P_1$	9.21	8.46	8.25	25.92	8.64
2	$P_1 \times P_2$	11.95	12.46	13.18	37.59	12.53
3	$P_1 \times P_3$	17.21	15.99	16.21	49.41	16.47
4	$P_1 \times P_4$	20.62	21.43	20.68	62.73	20.91
5	$P_1 \times P_5$	18.63	17.98	16.55	53.16	17.72
6	$P_1 \times P_6$	15.98	16.83	17.08	49.89	16.63
7	$P_2 \times P_1$	17.36	17.21	16.19	50.76	16.92
8	$P_2 \times P_2$	9.87	10.76	10.96	31.59	10.53
9	$P_2 \times P_3$	11.73	12.54	11.49	35.76	11.92
10	$P_2 \times P_4$	17.36	17.82	15.55	50.73	16.91
11	$P_2 \times P_5$	13.26	12.82	11.06	37.14	12.38
12	$P_2 \times P_6$	11.21	12.76	13.71	37.68	12.56
13	$P_3 \times P_1$	10.93	11.98	13.84	36.75	12.25
14	$P_3 \times P_2$	16.30	15.68	18.15	50.13	16.71
15	$P_3 \times P_3$	11.28	12.10	11.24	34.62	11.54
16	$P_3 \times P_4$	19.08	20.37	18.27	57.72	19.24
17	$P_3 \times P_5$	11.92	12.63	12.5	37.05	12.35
18	$P_3 \times P_6$	15.83	17.21	15.92	48.96	16.32
19	$P_4 \times P_1$	18.36	17.98	15.17	51.51	17.17
20	$P_4 \times P_2$	16.85	17.54	17.54	51.93	17.31
21	$P_4 \times P_3$	11.82	10.89	11.13	33.84	11.28
22	$P_4 \times P_4$	13.21	12.58	11.68	37.47	12.49
23	$P_4 \times P_5$	11.21	12.37	11.49	35.07	11.69
24	$P_4 \times P_6$	17.63	16.82	14.66	49.11	16.37
25	$P_5 \times P_1$	17.63	18.82	18.42	54.87	18.29
26	$P_5 \times P_2$	15.08	15.64	12.99	43.71	14.57
27	$P_5 \times P_3$	14.63	15.21	15.22	45.06	15.02
28	$P_5 \times P_4$	17.21	16.12	17.01	50.34	16.78
29	$P_5 \times P_5$	12.63	13.52	14.2	40.35	13.45
30	$P_5 \times P_6$	16.93	15.62	16.71	49.26	16.42
31	$P_6 \times P_1$	18.23	17.83	15.6	51.66	17.22
32	$P_6 \times P_2$	15.82	16.21	17.05	49.08	16.36
33	$P_6 \times P_3$	16.83	16.38	15.42	48.63	16.21

Contd...

Contd...

34	$P_6 \times P_4$	21.32	19.68	21.79	62.79	20.93
35	$P_6 \times P_5$	11.28	11.67	12.12	35.07	11.69
36	$P_6 \times P_6$	13.08	12.83	11.35	37.26	12.42
	Total	539.48	544.74	530.38	1614.60	14.95

Analysis of Variance for RBD (Diallel cross) :

Sources	df	SS	MS	F Ratio
Replicates	2	2.88	1.44	1.79
Treatments	35	952.28	27.21**	33.71
Parents	5	44.25	8.85**	10.97
Hybrids	29	653.95	22.55**	27.94
Parent Vs. Hybrids	1	254.08	254.08**	314.81
F1's	14	356.49	25.46**	31.55
Reciprocals	14	291.09	20.79**	25.76
F1 Vs Reciprocals	1	6.37	6.37**	7.89
Error	70	56.50	0.81	
Total	107	1011.66	9.45	

There are significant differences among the treatments (parents, direct and reciprocal crosses for ear head length, hence proceeded for combining ability analysis.

Prepare two way table

	P1	P2	P3	P4	P5	P6	Total
P1	8.64	12.53	16.47	20.91	17.72	16.63	92.90
P2	16.92	10.53	11.92	16.91	12.38	12.56	81.22
P3	12.25	16.71	11.54	19.24	12.35	16.32	88.41
P4	17.17	17.31	11.28	12.49	11.69	16.37	86.31
P5	18.29	14.57	15.02	16.78	13.45	16.42	94.53
P6	17.22	16.36	16.21	20.93	11.69	12.42	94.83
Total	90.49	88.01	82.44	107.26	79.28	90.72	538.2

$$\textbf{SS due to GCA} = \frac{1}{2P}[S(Yi. + Y.j)^2] - \frac{1}{2P}Y^2...$$

$$= \frac{1}{2 \times 6}[(92.90 + 90.49)^2 + ... + (94.83 + 90.72)^2]$$

$$-\frac{2}{36}(538.20)^2] = 38.53$$

SS due to SCA $= \frac{1}{2}\,[\Sigma\Sigma Yij\,(Yij + Yji) - \frac{1}{2P}\,\Sigma(Yi. + Yj.)^2) + \frac{1}{P^2}Y^2]$

$$= \frac{1}{2}[8.64\,(8.64 + 8.64) + 12.53\,(12.53 + 16.92) + \ldots + 12.42(12.42 + 12.42)]$$

$$- \frac{1}{2 \times 6}\,[(92.90 + 90.49)^2 + \ldots\ldots + (94.83 + 90.72)^2] + \frac{1}{6 \times 6}\,(538.2)^2]$$

$$= 159.70$$

SS due to RCA $= \frac{1}{2}[\Sigma\Sigma(Yij + Yji)^2]$

$$= \frac{1}{2}\,[(12.53 + 16.92)^2 + \ldots\ldots + (16.42 + 11.69)^2] = 116.83$$

Analysis of Variance for Combining Ability

Sources	df	SS	MSS	F
GCA	5	38.53	7.71** (Mg)	28.55
SCA	15	159.70	10.65** (Ms)	39.44
RCA	15	116.83	7.79** (Mr)	28.85
Error	70	56.50	0.27** (Me')	

Estimate the Genetic components

Model I

$$\sigma^2 GCA = \Sigma g_i^2 = \frac{Mg - Me'}{2P} = \frac{7.71 - 0.27}{12} = 0.62$$

$$\sigma^2 SCA = \Sigma g_i^2 = MS - Me' = 10.65 - 0.27 = 10.38$$

$$\sigma^2 RCA = \frac{(Mr - Me')}{2} = \frac{7.79 - 0.27}{2} = 3.76$$

The ratio of $\frac{\sigma^2 GCA}{\sigma^2 SCA} = \frac{0.62}{10.38} = 0.06$

Since the ratio of $\frac{\sigma^2 GCA}{\sigma^2 SCA}$ and $\frac{\sigma^2_A}{\sigma^2_D}$ less than one, the earhead length is governed by non additive gene action.

Model I

$$\sigma^2_A = 2\,\sigma^2 GCA = (2 \times 0.62) = 1.22$$

$$\sigma^2_D = \sigma^2 SCA = 10.38$$

The ratio of $\frac{\sigma^2_A}{\sigma^2_D} = 0.12$

The ratio of $\frac{\sigma^2_A}{\sigma^2_D}$ less than one, the earhead length is governed by non additive gene action.

Estimation of genetic Effects

(*i*) *GCA* effects of parents

$$g_i = \frac{1}{2P}(Y_{i.} - Y_{.j}) - \frac{1}{2P}Y$$

$$g_1 = \frac{1}{2 \times 6}(92.40 + 90.49) - \frac{1}{6 \times 6}(538.2) = 0.33$$

Similarly, the *gca* effects of all parents are estimated

Parent	GCA effect	*t* cal
P1	0.33*	2.427
P2	–0.85**	–6.186
P3	–0.71**	–5.201

Parent	gca effect	*t* cal
P4	1.18**	8.619
P5	–0.47**	–3.400
P6	0.51**	3.741

$$\text{SE for gca effect of parents} = \sqrt{\frac{(P-1)}{2P^2}Me'} = \left(\frac{5}{72} \times 0.27\right)^{\frac{1}{2}} = 0.137$$

$$\text{SE for } gi - gj = \sqrt{\frac{1}{P}Me'} = \left(\frac{1}{6} \times 0.27\right)^{1/2} = 0.21$$

t cal. $= \frac{gi}{SEgi}$, '*t*' table value at error *df* (70) is 1.99 at 5% and 2.63 at 1.% level of significance.

(*ii*) sca effect of hybrids (Direct crosses)

$$(Sij) = \frac{1}{2}(Y_{ij} + Y_{ji}) - \frac{1}{2P}(Y_i + Y_j + Y_i + Y_j) \cdot \frac{1}{P^2}Y$$

$$P_1 \times P_2 = \frac{1}{2}(12.53 + 16.92) - \frac{1}{2 \times 6}(92.90 + 90.49 + 81.22 + 88.01) + \frac{1}{6 \times 6}(538.2)$$
$$= 0.29$$

Similarly, the *sca* effects of all direct crosses are estimated

Cross	*SCA* effects	t cal.
$P_1 \times P_2$	0.29	0.97
$P_1 \times P_3$	–0.21	–0.70
$P_1 \times P_4$	2.58**	8.60
$P_1 \times P_5$	3.19**	10.63
$P_1 \times P_6$	1.13**	3.77
$P_2 \times P_3$	0.93**	3.10
$P_2 \times P_4$	1.83**	6.10
$P_2 \times P_5$	–0.16	–0.53

Cross	*SCA* effects	t cal.
$P_2 \times P_6$	–0.15	–0.50
$P_3 \times P_4$	–0.16	–0.53
$P_3 \times P_5$	–0.09	–0.30
$P_3 \times P_6$	1.52**	5.07
$P_4 \times P_5$	–1.43**	–4.77
$P_4 \times P_6$	2.01**	6.70
$P_5 \times P_6$	–0.94**	–3.13

$$\text{SE for SCA effect of hybrids (sij)} = \sqrt{\frac{(P^2 - 2p + 2)}{p^2} Me'}$$

$$= \left(\frac{1}{72} \times (36 - 12 + 2) \times 0.27\right)^{1/2}$$

$$= 0.30$$

$$\text{SE } (sij - sik) = \sqrt{\frac{(P-1)}{P} Me'}$$

$$= \left(\frac{5}{2} \times 0.27\right)^{1/2} = 0.82$$

(*iii*) RCA effects of hybrids (*rij*) $= \frac{1}{2}(Y_{ij} + Y_{ji})$

$$\text{For P2} \times \text{P1} = \frac{1}{2}(12.53 - 16.92) = -2.20$$

Similarly, the *rca* effects of all reciprocal crosses are estimated

Cross	RCA	t cal
$P_2 \times P_1$	–2.20**	–5.95
$P_3 \times P_1$	2.11**	5.70
$P_4 \times P_1$	1.87**	5.05
$P_5 \times P_1$	–0.29	–0.78
$P_6 \times P_1$	–0.30	–0.81
$P_3 \times P_2$	–2.40	–6.49
$P_4 \times P_2$	–0.20	–0.54
$P_5 \times P_2$	–1.10**	–2.97

Cross	RCA	t cal
$P_6 \times P_2$	–1.90**	–5.14
$P_4 \times P_3$	3.98**	10.76
$P_5 \times P_3$	–1.34**	–3.62
$P_6 \times P_3$	0.05	0.14
$P_5 \times P_4$	–2.55**	–6.89
$P_6 \times P_4$	–2.28**	–6.16
$P_6 \times P_5$	2.37**	6.41

$$\text{SE for } rca \text{ effect of hybrids } (rij) = \sqrt{\frac{1}{2} Me'} = (0.5 \times 0.27)^{1/2} = 0.37$$

$$\text{SE } (rij - rkl) = \sqrt{Me^2} = (0.27)^{1/2} = 0.52$$

CONCLUSION

The analysis of variance for combining ability revealed that variances due to GCA, SCA and RCA were highly significant for ear head length of pearl millet. This indicated that both additive and non additive gene actions were involved in the inheritance of this trait. However, the ratio of gca/sca variance indicated that non additive gene action played major role in the inheritance of ear head length.

The estimates of gca effects revealed that parents *viz.*, P_1, P_4, and P_6 had positive and significant gca effects for ear head length and hence found to be the best general combiners, whereas, the parents P_2, P_3 and P_5 were found to be poor combiners as they exhibited significant negative effect.

Seven straight and four reciprocal crosses showed significant high sca effects in desirable direction for ear head length. The cross combinations *viz.*, $P_1 \times P_4$, $P_1 \times P_6$ and $P_4 \times P_6$ (direct crosses) and $P_4 \times P_1$ (reciprocal cross) with significant sca effects for ear head length involved both the parents with high gca effects. Thus additive genetic variance was predominant and hence recurrent selection for gca or synthetic and composite breeding in pearl millet for these crosses will be effective for further genetic improvement.

High SCA for ear head length was recorded in the most of the crosses *viz.*, $P_1 \times P_5$, $P_2 \times P_4$, $P_3 \times P_6$, $P_3 \times P_1$, $P_4 \times P_3$ and $P_6 \times P_5$ involving one parent with high and other with low gca effect Indicating the presence of non additive gene action. Such crosses could produce promising segregants only if additive gene effects are important in general good combiners and complementary epistasis effects in F_1 and both act in same direction to maximize the expression of the trait. Therefore, recurrent selection for specific combining followed by pedigree breeding or biparental mating could be followed in segregating generation of the crosses.

ANALYSIS OF VARIANCE FOR COMBINING ABILITY–METHOD II

The formulae to work out degrees of freedom and sum of squares due to various sources of variation and for combining ability analysis in Method I are as below

Source	*df*	Sum of squares
GCA	$P-1$	$\frac{1}{2}[\Sigma(Yi. + Y.ii)^2 - \frac{4}{p^2}Y^2_{..}]$
SCA	$\frac{P(P-1)}{2}$	$\Sigma\Sigma Yij^2 - \frac{1}{P+2}[\Sigma(Yi. + Yii)^2 + \frac{2}{(P+1)(P+2)}Y^2]$
Error	$(r-1)(t-1)$	
Total	$\frac{P(P+1)}{2}-1$	

Source	MS	Expectation of Mean Squares	
		Model I	Model II
GCA	Mg	$\sigma_e^2 + P + 2\frac{1}{p-1}\Sigma g_i^2$	$\sigma_e^2 + \sigma_s^2 + (p+2)\sigma_g^2$
SCA	Ms	$\sigma_e^2 + \frac{2}{p(p-1)}\Sigma\Sigma S_{ij}^2$	$\sigma_e^2 + \sigma_s^2$
Error	Me'	σ_e^2	σ_e^2

$$Me' = \frac{EMS}{r} \text{ he EMS is from analysis of variance of RBD}$$

GENETIC COMPONENTS

Model I

$$\text{Variance due to GCA} = \frac{1}{P-1} = \Sigma g_i^2 = \frac{Mg - Me'}{2P}$$

$$\text{Variance due to SCA} = \frac{2}{P(P-1)} = \Sigma\Sigma S_{ij}^2 = Mg - Me'$$

The ratio of GCA variance/SCA variance indicates the relative significance of additive versus non– additive genetic variance

Model II

$$\text{Variance due to GCA} = \sigma^2_g = \frac{1}{P+2}(Mg - Ms)$$

$$\text{Variance due to SCA} = \sigma_S^2 = MS - Me'$$

(*iii*) F = 1

$\sigma^2_A = \sigma^2_g$ and $\sigma^2_D = \sigma^2_S$

(*iv*) F = 0

$\sigma^2_A = 4\sigma^2_g$ and $\sigma^2_D = 4\sigma^2_S$

The ratio of σ^2_A/σ^2_D indicates the relative significance of additive versus non–additive genetic variance.

COMBINING ABILITY EFFECTS

$$\text{gca effects of parents } (g_i) = \frac{1}{P+2}[(Y_{i.} + Y_{ii..}) - \frac{2}{P}Y..]$$

$$\text{sca effects of hybrids } (Sij) = Y_{ij} - \frac{1}{P+2} - (Y_i + Y_{ii} + Y_j + Y_{jj}) + \frac{2}{(P+1)(P+2)}Y..$$

Where,

$Y_{..}$ – Grand total of all hybrids

Y_i – Row total of ith parent

Y_{ii} – Mean value of ith parent

$Y.j_.$ – Column total of ith parent

$Y_{jj.}$ – Mean value of of jth parent

Y_{ij} – Mean value of ij[th] hybrid

P – Number of parents involved

Significance of effects is tested by '*t*' test

$$\text{SE for gca effect of parents} = \sqrt{\frac{(p-1)}{p(P+2)}Me'}$$

$$\text{SE for sca effect of hybrids } (sij) = \sqrt{\frac{P(P-1)}{(P+1)(P+2)} Me'}$$

$$\text{SE for } (sij - sik) = \sqrt{\frac{2(P+1)}{(P+2)} Me'}$$

$$\text{SE for } (sij - skl) = \sqrt{\frac{2(P+1)}{(P+2)} Me'}$$

$'t' = \frac{\text{Effect}}{\text{SE}}$ The calculated '*t*' value is compared with '*t*' table at error *df*

SOLVED EXERCISE

Diallel Analysis –method II

Seven pure lines of chickpea were crossed in diallel fashion following Method II. The data on seed yield/plant (g) was measured in three replication for 21 genotypes of Chickpea (15F_1s and 7 parents). The analysis is done as follows.

Genotype	Seed yield /plant (g)				
	RI	RII	RIII	Total	Mean
$P_1 \times P_1$	21.04	22.22	23.85	67.11	22.37
$P_1 \times P_2$	21.76	18.80	18.29	58.85	19.62
$P_1 \times P_3$	19.95	21.26	22.68	63.89	21.30
$P_1 \times P_4$	20.73	21.81	20.25	62.79	20.93
$P_1 \times P_5$	25.55	21.86	26.97	74.38	24.79
$P_1 \times P_6$	25.81	23.23	24.32	73.36	24.45
$P_1 \times P_7$	31.09	25.88	28.92	85.89	28.63
$P_2 \times P_2$	23.60	26.02	22.39	72.01	24.00
$P_2 \times P_3$	18.97	21.13	22.70	62.80	20.93
$P_2 \times P_4$	22.07	21.60	19.27	62.94	20.98
$P_2 \times P_5$	14.56	13.10	12.06	39.72	13.24
$P_2 \times P_6$	23.20	21.40	25.77	70.37	23.46
$P_2 \times P_7$	23.82	25.17	23.02	72.01	24.00
$P_3 \times P_3$	24.60	26.32	22.29	73.21	24.40
$P_3 \times P_4$	23.16	20.29	18.01	61.46	20.49
$P_3 \times P_5$	23.95	30.04	27.33	81.32	27.11
$P_3 \times P_6$	25.99	27.76	28.75	82.50	27.50
$P_3 \times P_7$	28.58	28.11	24.46	81.15	27.05

Contd...

Contd...

$P_4 \times P_4$	24.41	26.82	22.28	73.51	24.50
$P_4 \times P_5$	26.97	28.84	27.21	83.02	27.67
$P_4 \times P_6$	19.89	26.74	20.52	67.15	22.38
$P_4 \times P_7$	21.81	20.23	24.67	66.71	22.24
$P_5 \times P_5$	27.44	26.66	22.11	76.21	25.40
$P_5 \times P_6$	32.91	35.45	30.92	99.28	33.09
$P_5 \times P_7$	35.56	37.20	32.15	104.91	34.97
$P_6 \times P_6$	26.74	22.52	24.15	73.41	24.47
$P_6 \times P_7$	25.90	19.12	23.10	68.12	22.71
$P_7 \times P_7$	17.57	15.79	16.07	49.43	16.48
Total	677.63	675.37	654.51	2007.51	23.90

Analysis of Variance for RBD (diallel cross) :

Source	**df**	**SS**	**MS**	**F Ratio**
Replicates	2	11.60	5.80	1.26
Treatments	27	1565.58	57.98**	12.60
Parents	6	168.20	28.03**	6.09
Hybrids	20	1379.06	68.95**	14.98
Parent Vs.Hybrids	1	18.32	18.32	3.98
Error	54	248.53	4.60	
Total	83	1825.71		

There are significant differences among the treatments (parents and hybrids) for seed yield/plant, hence proceed for combining ability analysis.

Prepare two way table

	P1	**P2**	**P3**	**P4**	**P5**	**P6**	**P7**	**Total**	**Yi.+Yii**
P1	22.37	19.62	21.30	20.93	24.79	24.45	28.63	162.09	184.46
P2		24.00	20.93	20.98	13.24	23.46	24.00	126.62	170.24
P3			24.40	20.49	27.11	27.50	27.05	126.55	193.18
P4				24.50	27.67	22.38	22.24	96.80	183.70
P5					25.40	33.09	34.97	93.47	211.68
P6						24.47	22.71	47.18	202.53
P7							16.48	16.48	192.56
Total								669.17	

Yi. = Row total + value of reciprocal crosses

$$\text{SS due to GCA} = \frac{1}{P+2}[\Sigma(Yi. + Yii)^2 - \frac{4}{P}Y^2..]$$

$$= \frac{1}{9}((184.46)^2 + ... + (192.56)^2 - \frac{1}{7}(669.17)^2 = 121.66$$

$$\text{SS due to SCA} = \Sigma\Sigma Yij^2 - \frac{1}{P+2}[\Sigma(Yi + Yii)^2 + \frac{2}{(P+1)(P+2)}Y^2]$$

$$= (22.37)^2 + (19.62)^2 + (21.30)^2 + ... + (16.48)^2 - \frac{1}{9}[(184.46)^2 + ... + (192.56)^2 + (669.17)^2 = 400.20$$

Analysis of Variance for Combining Ability

Source	df	SS	MSS	F ratio
GCA	6	121.66	20.28** (Mg)	13.22
SCA	21	400.20	19.06** (Ms)	12.42
Error	54	82.84	1.53 (Me')	

Estimation of the Genetic Components

Model I

$$\text{Variance due to } \sigma^2\text{gca} = \frac{\text{Mg} - \text{Me'}}{P+2} = \frac{20.28 - 1.53}{7+2} = 2.08$$

$$\text{Variance due to } \sigma^2\text{sca} = \text{Ms} - \text{Me'} = 19.06 - 1.53 = 17.53$$

$$\text{The ratio of } \frac{\sigma^2\text{gca}}{\sigma^2\text{sca}} = \frac{2.08}{17.53} = 0.12$$

The ratio of $\frac{\sigma^2\text{gca}}{\sigma^2\text{sca}}$ is less than 1, which indicates that the seed yield is governed by non additive gene action.

Model II

$$\text{Variance due to } \sigma^2\text{gca} = \frac{\text{Mg} - \text{Ms}}{P+2} = \frac{20.28 - 19.06}{7+2} = 0.12$$

$$\text{Variance due to } \sigma^2\text{sca} = \text{Ms} - \text{Me'} = 19.06 - 1.53 = 17.53$$

$$\sigma^2_A = 2\sigma^2\text{gca} = (2 \times 0.12) = 0.24$$

$$\sigma^2_D = \sigma^2\text{sca} = 17.53$$

$$\text{The ratio of } \frac{\sigma^2_A}{\sigma^2_D} = \frac{0.28}{17.53} = 0.01$$

The ratio of σ^2gca/σ^2sca and σ^2_A / σ^2_D is less than 1, which indicates that the seed yield is governed by non additive gene action.

Estimation of Genetic Effects

$$\textit{GCA} \textbf{ effects of parents } (g_i) = \frac{1}{P+2}(Y_{i.} + Y_{ii}) - \frac{2}{P}Y..]$$

$$\text{for } P_1, g_i = \frac{1}{7+2} = [(184.46 - \frac{2}{7}(669.17)] - 0.75$$

Similarly, the *GCA* effects of all parents are estimated

Parent	*gca* effect	*t* cal
P1	–0.75	–1.97
P2	–2.33**	–6.13
P3	0.22	0.58
P4	–0.83**	–2.18

Parent	*gca* effect	t cal
P5	2.28**	6.00
P6	1.26**	3.32
P7	0.15	0.39

SE for gca effect of parents = $\sqrt{\frac{(p-1)}{p(P+2)}Me'} = \left(\frac{6}{63} \times 1.53\right)^{1/2} = 0.38$

SE for $gi - gj$ = $\sqrt{\frac{2Me'}{P+2}} = \left(\frac{2 \times 1.53}{9}\right)^{\frac{1}{2}} = 0.58$

t cal. = $\frac{gi}{SEgi}$, '*t*' table value at error *df* (54) is 2.00 at 5 % and 2.67 at 1.% level of significance.

(ii) *sca* effects of hybrids (*sij*) = $Yij \frac{1}{P+2}(Y_i + Y_{ii} + Y_j + Y_{jj}) + \frac{2}{(P+1)(P+2)} Y..$

For $P_1 \times P_2 = 19.62 - \frac{1}{7+2}(184.46 + 170.24) + \frac{1}{8 \times 9}(669.17) = -1.21$

Similarly, the *SCA* effects of all crosses are estimated

Cross	*sca* effect	t cal.		Cross	*sca* effect	t cal.		Cross	*sca* effect	t cal.
$P_1 \times P_2$	–1.21	–1.09		$P_2 \times P_4$	0.24	0.22		$P_3 \times P_7$	2.78*	2.50
$P_1 \times P_3$	–2.08	–1.87		$P_2 \times P_5$	–10.61**	–9.56		$P_4 \times P_5$	2.33*	2.10
$P_1 \times P_4$	–1.39	–1.25		$P_2 \times P_6$	0.63	0.56		$P_4 \times P_6$	–1.94	–1.75
$P_1 \times P_5$	–0.63	–0.57		$P_2 \times P_7$	2.28*	2.05		$P_4 \times P_7$	–0.98	–0.88
$P_1 \times P_6$	0.04	0.04		$P_3 \times P_4$	–2.80*	–2.52		$P_5 \times P_6$	5.66**	5.10
$P_1 \times P_7$	5.33**	4.80		$P_3 \times P_5$	0.71	0.64		$P_5 \times P_7$	8.64**	7.79
$P_2 \times P_3$	–0.86	–0.77		$P_3 \times P_6$	2.12	1.91		$P_6 \times P_7$	–2.60	–2.35

SE for sca effect of hybrids (*sij*) = $\sqrt{\frac{2(P-1)}{(P+1)(P+2)}Me'} = (1.23)^{1/2} = 1.11$

SE for (*sij*–*sik*) = $\sqrt{\frac{2(P+1)}{(P+2)}Me'} = (2.72)^{1/2} = 1.65$

SE for (*sij*–*skl*) = $\sqrt{\frac{2(P+1)}{(P+2)}Me'} = (2.72)^{1/2} = 1.65$

t cal. = $\frac{gi}{SEgi}$, 't' table value at error *df* (54) is 2.00 at 5 % and 2.67at 1.% level of significance.

CONCLUSION

The analysis of variance for combining ability revealed that variances due to gca and sca were highly significant for seed yield of chickpea. This indicated that both additive and non additive gene actions were involved in the inheritance of this trait. However, the ratio of additive to the dominance variance indicated that non additive gene action played major role in the inheritance of seed yield.

The estimates of *gca* effects revealed that parents viz., P_5, and P_6 had positive and significant *gca* effects for seed yield and hence found to be the best general combiners.

Six crosses showed significant high sca effects in desirable direction for grain yield. The cross combination viz., $P_5 \times P_6$ with positive and significant *sca* effects for grain yield involved both the parents with high gca effects. Thus additive genetic variance was predominant and hence simple selection by pedigree method to develop pure line variety through this cross will be effective for further genetic improvement.

High *sca* for grain yield was also recorded in the crosses *viz.,* $P_5 \times P_7$, $P_4 \times P_5$ involving one parent with high and other with low *gca* effect indicating presence of non additive gene action. Therefore, recurrent selection method of breeding could be followed.

The crosses *viz.,* $P_1 \times P_7$ and $P_3 \times P_7$ involving L × L combiners exhibited high sca effects. This might have arisen due to the more genetic diversity among the parents involved. For effective utilization of these cross combinations, it is suggested to make inter se crosses among these combinations so as to have a multiple parents input in a central gene pool which will supplement genetic recombination and will break undesirable linkages.

ANALYSIS OF VARIANCE FOR COMBINING ABILITY

Method III

In method III, both F_1s and reciprocals crosses constitutes the experimental material. The total entries being P (P–1).

The formulae to work out degrees of freedom and sum of squares due to various sources of variation and for combining ability analysis in Method III are as below

Source	*df*	Sum of squares
GCA	$P-1$	$\frac{1}{2(P-2)}[\Sigma(Yi. + Y.j)^2] - \frac{1}{P(P-2)}Y^2..$
SCA	$\frac{P(P-3)}{2}$	$\frac{1}{2}\Sigma\Sigma(Yij + Yji)^2 - \frac{1}{P(P-2)}\Sigma(Yi. + Y.j)^2 + \frac{1}{(P-1)(P-2)}Y^2..$
Reciprocal	$\frac{P(P-1)}{2}$	$\frac{1}{2}[\Sigma\Sigma(Yij + Yji)^2]$
Error	$(r-1)(t-1)$	
Total	$P(P-1)-1$	

Source	MS	Expectation of Mean Squares	
		Model I	Model II
GCA	Mg	$\sigma_e^2 + 2(P-2)\frac{1}{p-1}\Sigma g_i^2$	$\sigma_e^2 + 2\sigma_g^2 + 2(p-2)\sigma_g^2$
SCA	Ms	$\sigma_e^2 + 2\frac{1}{p(p-3)}\Sigma\Sigma S_{ij}^2$	$\sigma_e^2 + 2\sigma_s^2$
RCA	Mr	$\sigma_e^2 + 2\frac{2}{p(p-1)}\Sigma\Sigma r_{ij}^2$	$\sigma_e^2 + 2\sigma_r^2$
Error	Me′	σ_e^2	σ_e^2

$Me' = \frac{EMS}{r}$ The EMS is from analysis of variance of RBD

Genetic Components

Model I

$$\text{Variance due to GCA} = \frac{1}{P-1} = \Sigma g_i^2 = \frac{Mg - Me'}{2(P-2)}$$

$$\text{Variance due to SCA} = \frac{2}{P(P-3)} = \Sigma\Sigma S_{ij}^2 = \frac{Ms - Me'}{2}$$

The ratio of GCA variance/SCA variance indicates the relative significance of additive versus non– additive genetic variance.

Model II

$$\text{Variance due to GCA} = \sigma^2_g = \frac{1}{2(P-2)} Mg - Ms$$

$$\text{Variance due to SCA} = \sigma^2_S = \frac{Ms - Me'}{2}$$

(*i*) F = 1

$\sigma^2_A = \sigma^2_g$ and $\sigma^2_D = \sigma^2_S$

(*ii*) $F = 0^2$

$\sigma^2_A = 4\sigma^2_g$ and $\sigma^2_D = 4\sigma^2_S$

The ratio of σ^2_A/σ^2_D indicates the relative significance of additive versus non–additive genetic variance.

Combining Ability Effects

Method III

$$\text{gca effects of parents } (g_i) = \frac{1}{2P(P-2)}[P(Y_{i.} + Y_{.j}) - 2Y_{..}]$$

$$\text{sca effects of hybrids } (Sij) = \frac{1}{2}(Y_{ij} + Y_{ji}) - \frac{1}{2(P-2)}(Y_{i.} + Y_{.j} + Y_{i.} + Y_{.j}) + \frac{2}{(P+1)(P+2)} Y_{..}$$

$$rca \text{ effects of hybrids } (rij) = \frac{1}{2}(Y_{ij} + Y_{ji})$$

Significance of effects is tested by '*t*' test

$$\text{SE for gca effect of parents} = \sqrt{\frac{(p-1)}{2P(P-2)} Me'}$$

$$\text{SE for sca effect of hybrids} = \sqrt{\frac{p-3}{2(P-1)} Me'}$$

$$\text{SE for rca effect of hybrids} = \sqrt{\frac{1}{2} Me'}$$

$$\text{SE } (g_1 - g_j) = \sqrt{\frac{1}{(P-2)} Me'}$$

$$\text{SE } (s_{ij} - s_{ik}) = \sqrt{\frac{(P-3)}{(P-2)} Me'}$$

$$\text{SE } (s_{ij} - s_{kl}) = \sqrt{\frac{(P-4)}{(P-2)} Me'}$$

$'t' = \frac{\text{Effect}}{\text{SE}}$ The calculated '*t*' value is compared with '*t*' table at error df.

SOLVED EXERCISE

Diallel Analysis Griffing (1956) method III

The data measured on the character ear head length in pearl millet and given in Table of diallel analysis of Method I . After ignoring the parental values (the diagonal value) is considered herewith the analysis by following Griffings Method III..

Analysis of Variance for RBD

Source	df	SS	MS	F Ratio
Replicates	2	2.37	1.19	1.33
Genotypes	29	645.56	22.26**	25.01
Error	58	51.61	0.89	
Total	89	699.54		

There are significant differences among the genotypes for ear head length, hence proceeded for combining ability analysis.

Prepare two way table

	P1	P2	P3	P4	P5	P6	Total
P1	–	12.53	16.47	20.91	17.72	16.63	84.26
P2	16.92	–	11.92	16.91	12.38	12.56	70.69
P3	12.25	16.71	–	19.24	12.35	16.32	76.87
P4	17.17	17.31	11.28	–	11.69	16.37	73.82
P5	18.29	14.57	15.02	16.78	–	16.42	81.08
P6	17.22	16.36	16.21	20.93	11.69	–	82.41
Total	81.85	77.48	70.90	94.77	65.83	78.30	469.13

SS due to GCA $= \frac{1}{2(P-2)} [\Sigma(Yi. + Y.j)^2] - \frac{1}{P(P-2)} Y^{2}..$

$= \frac{1}{2 \times 4} [(84.26 + 81.85)2 + ... + (82.41 + 78.30)2]$

$- \frac{2}{24} (469.13)2] = 61.72$

SS due to SCA $= \frac{1}{2} \Sigma\Sigma(Yij + Yji)^2 - \frac{1}{2(P-2)} \Sigma(Yi + Yj)^2) + \frac{1}{(P-1)(P-2)} Y^{2}..$

$= \frac{1}{2}[(12.53 + 16.92)^2 + ...(16.37 + 20.93)^2 - \frac{1}{2 \times 4} [(84.26 + 81.25)^2 + ...+$

$(82.41 + 78.3)2] + \frac{1}{5 \times 4} (469.13)^2 = 36.64$

SS due to RCA $= \frac{1}{2}[\Sigma\Sigma(Yij + Yji)^2]$

$= \frac{1}{2} [(12.53 + 16.92)^2 + ...+(16.42 + 11.69)^2] = 116.83$

Analysis of Variance for Combining Ability

Source	*df*	SS	MSS	F
GCA	5	61.72	12.34** (Mg)	41.14
SCA	9	36.64	4.07** (Ms)	13.57
RCA	15	116.83	7.79** (Mr)	25.96
Error	58		0.30 (Me′)	

Estimate the Genetic components

Model I

$\sigma^2 gca = \ = = \ = 1.03$

$\sigma^2 sca \ = = \ 4.07 - 0.30 = 1.88$

$\sigma^2 rca \ = = \ 7.79 - 0.30 = 3.74$

The ratio of σ^2gca/ σ^2sca = 1.03/1.88 = 0.55

Since the ratio of σ^2gca/ σ2sca is less than one, the earhead length is governed by non additive gene action.

Model II

Variance due to GCA = $\sigma_9^2 = \frac{1}{2(P-2)}$ = = Mg–Ms = (1/10) × (12.34 –4.07)= 0.83

Variance due to SCA = $\sigma_5^2 = \frac{M_S - M_e^1}{2}$ = 1.88

σ^2_A = 2 σ^2gca = (2 × 0.83)= 1.66

σ^2_D = σ^2sca = 1.88

The ratio of σ^2_A / σ^2_D = 0.88

Since the ratio of σ^2_A / σ^2_D is less than one, the earhead length is governed by non additive gene action.

Estimation of genetic Effects

***gca* effects of parents**

$$g_i = \frac{1}{2P(P-2)}[P(Y_{i.} + Y.j) - Y..]$$

$$g_1 = [6(84.26 + 81.85) - 2 \times (469.16) = 1.22$$

Similarly, the *gca* effects of all parents are estimated

Parent	gca effect	t cal
P1	1.22**	6.76
P2	–1.103**	–5.70
P3	–1.08**	–5.98

Parent	gca effect	t cal
P4	1.53**	8.48
P5	–1.18**	–6.57
P6	0.54**	3.01

$$\text{SE for gca effect of parents} = \sqrt{\frac{(P-1)}{2P(P-2)}Me'} = \left(\frac{5}{48}\times 0.30\right)^{1/2} = 0.18$$

$$\text{SE for gi} - \text{gj} = \sqrt{\frac{1}{(P-2)}Me'} = \left(\frac{1}{4}\times 0.30\right)^{1/2} = 24$$

t cal. = $\frac{gi}{Segi}$ 't' table value at error df (58) is 2.00 at 5 % and 2.66 at 1.% level of significance.

sca effect of hybrids (Direct crosses)

$$(Sij) = \frac{1}{2}(Y_{ij} - Y_{ji}) - \frac{1}{2(P-2)}(Y_{i.} + Y_{,j} + Y_{.j} + Y_{.i}) + \frac{1}{(P-1)(p-2)}Y..$$

$$P_1 \times P_2 = \frac{1}{2}(12.53 + 16.92) - \frac{1}{2\times 4}(84.26 + 70.69 + 81.85 + 77.48) + \frac{1}{5\times 4}(469.13)$$

Similarly, the *sca* effects of all direct crosses are estimated

Cross	*sca* effects	t cal.
$P_1 \times P_2$	–1.10**	–3.68
$P_1 \times P_3$	–1.34**	–4.48
$P_1 \times P_4$	0.73*	2.45
$P_1 \times P_5$	2.41**	8.03
$P_1 \times P_6$	–0.40	–1.32
$P_2 \times P_3$	0.78*	2.60
$P_2 \times P_4$	0.97**	3.24
$P_2 \times P_5$	0.05	0.16

Cross	sca effects	t cal.
$P_2 \times P_6$	–0.69*	–2.31
$P_3 \times P_4$	–0.83**	–2.76
$P_3 \times P_5$	0.31	1.02
$P_3 \times P_6$	1.16**	3.87
$P_4 \times P_5$	–1.75**	–5.82
$P_4 \times P_6$	0.94**	3.15
$P_5 \times P_6$	–0.94**	–3.14

SE for sca effect of hybrids (Sij) = $\sqrt{\frac{(P-3)}{2(P-1)}Me'} = \sqrt{\frac{3}{2\times5}\times0.30} = 0.30$

$SE\ (s_{ij}-s_{ik}) = \sqrt{\frac{(P-3)}{(P-2)}Me'} = \sqrt{\frac{3}{4}\times0.30} = 0.47$

$SE\ (s_{ij}-s_{kl}) = \sqrt{\frac{(P-3)}{(P-2)}Me'} = \sqrt{\frac{2}{4}\times0.30} = 0.39$

(iii). **rca effects of hybrids (r*ij*)** = $\frac{1}{2}(Y_{ij}-Y_{ji})$

For $P_2 \times P_1$ = (12.53–16.92) = – 2.20

Similarly, the *rca* effects of all reciprocal crosses are estimated

Cross	rca effect	t cal
$P_2 \times P_1$	–2.20**	–5.63
$P_3 \times P_1$	2.11**	5.41
$P_4 \times P_1$	1.87**	4.79
$P_5 \times P_1$	–0.29	–0.73
$P_6 \times P_1$	–0.30	–0.76
$P_3 \times P_2$	–2.40**	–6.14
$P_4 \times P_2$	–0.20	–0.51
$P_5 \times P_2$	–1.10**	–2.81

Cross	rca effect	t cal
$P_6 \times P_2$	–1.90**	–4.87
$P_4 \times P_3$	3.98**	10.21
$P_5 \times P_3$	–1.34**	–3.42
$P_6 \times P_3$	0.05	0.14
$P_5 \times P_4$	–2.55**	–6.53
$P_6 \times P_4$	–2.28**	–5.85
$P_6 \times P_5$	2.37**	6.06

SE for rca effect of hybrids = $\sqrt{\frac{1}{2}Me'} = (0.5 \times 0.30)^{1/2} = 0.39$

$SE\ (r_{ij}-r_{kl}) = \sqrt{Me'} = (0.30)^{1/2} = 0.55$

CONCLUSION

The analysis of variance for combining ability revealed that variances due to GCA, SCA and RCA were highly significant for ear head length of pearl millet.

This indicated that both additive and non additive gene actions were involved in the inheritance of this trait. However, the ratio of GCA /SCA variance indicated that non–additive gene action played major role in the inheritance of ear head length.

The estimates of gca effects revealed that parents *viz.*, P_1, P_4, and P_6 had positive and significant gca effects for ear head length and hence found to be the best general combiners, whereas, the parents P_2, P_3 and P_5 were found to be poor combiners as they exhibited significant negative effect.

Six straight and three reciprocal crosses showed significant high sca effects in desirable direction for ear head length. The cross combinations *viz.*, $P_1 \times P_4$, and $P_4 \times P_6$ (direct crosses) with significant sca effects for ear head length involved both the parents with high gca effects. Thus additive genetic variance was predominant and hence recurrent selection for gca or synthetic and composite breeding in pearl millet for these crosses will be effective for further genetic improvement.

High sca for ear head length was recorded in the most of the crosses viz., $P_1 \times P_5$, $P_2 \times P_4$, $P_3 \times P_6$, $P_4 \times P_1$ and $P_4 \times P_3$ involving one parent with high and other with low gca effect indicating the presence of non additive gene action. Such crosses could produce promising segregants only if additive gene effects are important in general good combiners and complementary epistasis effects in F_1 and both act in same direction to maximize the expression of the trait. Therefore, recurrent selection for specific combining followed by pedigree breeding or biparental mating could be followed in segregating generation of the crosses.

Analysis of variance for combining ability– Method IV

In method IV, only F_1 crosses constitutes the experimental material. The total entries being P (P–1)/2.

The formulae to work out degrees of freedom and sum of squares due to various sources of variation and for combining ability analysis in Method IV are as below

Source	df	Sum of squares
GCA	P–1	$\frac{1}{P-2}\Sigma Yi^2_{.} - \frac{4}{P(P-2)}Y^2_{..}$
SCA	$\frac{P(P-3)}{2}$	$\Sigma\Sigma Yij^2 - \frac{1}{P-2}\,\Sigma Yi^2_{.} + \frac{2}{(P-1)(P-2)}Y^2_{..}$
Error	(*r*–1) (*t*–1)	
Total	$\frac{P(P-1)}{2}-1$	

Source	MS	Expectation of Mean Squares	
		Model I	Model II
GCA	Mg	$\sigma_e^2 + (P-2)\left(\frac{1}{p-1}\Sigma g_i^2\right)$	$\sigma_e^2 + \sigma_S^2 + (P22)\,\sigma_g^2$
SCA	Ms	$\sigma_e^2 + \left(\frac{2}{p(p-3)}\right)\Sigma\Sigma S_{ij}^2$	$\sigma_e^2 + \sigma_S^2$
Error	Me′	σ_e^2	σ_e^2

Me′ = . The EMS is from analysis of variance of RBD

Genetic Components

Model I

Variance due to GCA = $\frac{1}{P-1}\Sigma g_i^2 = \frac{Mg - Me'}{P-2}$

Variance due to SCA = $\frac{2}{P(P-3)}\Sigma\Sigma S_{ij}^2 = Ms - Me'$

The ratio of GCA variance/SCA variance indicates the relative significance of additive versus non– additive genetic variance

Model II

Variance due to GCA = $\sigma_g^2 = \frac{1}{P-2}Mg - Ms$

Variance due to SCA = $\sigma_S^2 = MS - Me'$

(v) F = 1

$\sigma_A^2 = 2\sigma_g^2$ and $\sigma_D^2 = \sigma_S^2$

(vi) F=0

$\sigma_A^2 = 4\sigma_g^2$ and $\sigma_D^2 = 4\sigma_S^2$

The ratio of σ_D^2/σ_H^2 indicates the relative significance of additive versus non–additive genetic variance.

Combining Ability Effects

gca effects of parents (g_i) = $\frac{1}{P(P-2)}(PY_i. - Y..)$

sca effects of hybrids (*Sij*) = $Yij - \frac{1}{P-2}(Y_j + Y._j) + \frac{2}{(P-1)(P-2)}Y..$

Significance of effects is tested by 't' test

SE for gca effect of parents = $\sqrt{\frac{P-1}{P(P-2)}Me'}$

SE for sca effect of hybrids = $\sqrt{\frac{(P-3)}{(P-1)}Me'}$

$$SE(g_i - g_j) = \sqrt{\frac{2}{(P-2)} Me'}$$

$$SE(s_{ij} - s_{ik}) = \sqrt{\frac{2(P-3)}{(P-2)} Me'}$$

$$SE(s_{ij} - s_{kl}) = \sqrt{\frac{2(P-4)}{(P-2)} Me'}$$

$'t' = \frac{\text{Effect}}{\text{SE}}$ The calculated 't' value is compared with 't' table at error df

SOLVED EXERCISE

Diallel Analysis Griffing (1956) – Method IV

The data measured on the character seed yield in chickpea and given in Table of diallel analysis of Method II . After ignoring the parental values (the diagonal value) is considered herewith the analysis by following Griffings Method IV

Analysis of Variance for RBD

Source	df	SS	MS	F Ratio
Replicates	2	2.96	1.48	0.30
Genotypes	20	1379.06	68.95**	13.76
Error	40	200.39	5.01	
Total	62	1582.42		

There are significant differences among the crosses for seed yield, hence proceed for combining ability analysis.

Prepare two way table

	P1	P2	P3	P4	P5	P6	P7	Total	Yi.
P1	–	19.62	21.30	20.93	24.79	24.45	28.63	139.72	139.72
P2		–	20.93	20.98	13.24	23.46	24.00	102.61	122.23
P3			–	20.49	27.11	27.50	27.05	102.15	144.38
P4				–	27.67	22.38	22.24	72.29	134.69
P5					–	33.09	34.97	68.06	160.87
P6						–	22.71	22.71	153.59
P7							–	–	159.6
Total								507.54	–

Yi. = Row total + value of reciprocal crosses

$$\text{SS due to GCA} = \frac{1}{P-2}\Sigma Yi.^2 - \frac{4}{P(P-2)}Y..^2$$

$$= \frac{1}{5}(139.72)^2 + ... + (159.60)^2 - \frac{1}{7\times5}(507.74)^2 = 311.95$$

$$\text{SS due to SCA} = \Sigma\Sigma Yij^2 - \frac{1}{P-2}\Sigma Yi.^2 + \frac{2}{(P-1)(P-2)}Y^2_{..}$$

$$= (19.62)^2 + (21.30)^2 + \text{—} + (22.71)^2 - \frac{1}{5}[(139.72)^2] + \ldots\ldots +$$

$$(159.60)^2 + \frac{2}{6\times5}(507.54)^2 = 307.05$$

Analysis of Variance for Combining Ability

Source	df	SS	MSS	F ratio
GCA	6	311.95	51.99** (Mg)	31.13
SCA	14	307.06	21.93** (Ms)	13.13
Error	40	–	1.67 (Me′)	

Estimation of the Genetic components

Model I

Variance due to $\sigma^2 gca = \frac{Mg-Me'}{P-2} = \frac{51.99-1.67}{7-2} = 10.06$

Variance due to $\sigma^2 sca = Ms - Me' = 21.93 - 1.67 = 20.26$

The ratio of $\sigma^2 gca/\sigma 2sca = 10.06/20.26 = 0.50$

The ratio of $\sigma^2 gca/\sigma^2 sca$ is less than 1, which indicates that the seed yield is governed by non additive gene action.

Model II

Variance due to GCA = $\sigma_g^2 = \frac{1}{P-2} Mg - Ms = 6.01$

Variance due to SCA = $\sigma_S^2 = Ms - Me' = 1.67$

$\sigma^2_A = 2\sigma^2 gca = (2 \times 6.01) = 12.02$

$\sigma^2_D = \sigma^2 sca = 20.26$

The ratio of $\sigma^2_A/\sigma^2_D = 12.02/20.26 = 0.59$

The ratio of σ^2_A/σ^2_D is less than 1, which indicates that the seed yield is governed by non additive gene action.

Estimation of genetic effects

(*ii*) *GCA* effects of parents (g_i) = $\frac{1}{P(P-2)}(PY_i - 2Y..)$

$$\text{for } P_1, g_i = \frac{1}{7\times5}[7(139.72) - 2X(507.54)] = -1.06$$

Similarly, the *gca* effects of all parents are estimated

Parent	gca effect	t cal
P1	–1.06	–2.00
P2	–4.56**	–8.60
P3	–0.13	–0.24
P4	–2.06**	–3.89

Parent	gca effect	t cal
P5	3.17**	5.98
P6	1.72**	3.24
P7	2.92**	5.51

SE for gca effect of parents = $\sqrt{\dfrac{(p-1)}{p(P-2)}Me'} = ((6/35)\times 1.67)^{1/2} = 0.53$

SE for gi –gj = $\sqrt{\dfrac{2Me'}{P-2}} = ((2\times 11.67)/5)^{1/2} = 0.82$

t cal. = $\dfrac{gi}{SEgi}$, 't' table value at error df (40) is 2.02 at 5 % and 2.70 at 1.% level of significance.

(ii) *SCA* effects of hybrids (*Sij*) = $Yij - \dfrac{1}{P-2}(Y_{i.} + Y_{.j}) + \dfrac{2}{(P-1)(P-2)}Y$

For $P_1 \times P_2 = 19.62 - \dfrac{1}{7-2}(139.72 + 122.73) + \dfrac{1}{6\times 5}(507.54) = -1.21$

Similarly, the *sca* effects of all crosses are estimated

Cross	sca effect	t cal.	Cross	sca effect	t cal.	Cross	sca effect	t cal.
$P_1 \times P_2$	1.07	1.01	$P_2 \times P_4$	3.43**	3.24	$P_3 \times P_7$	0.09	0.08
$P_1 \times P_3$	–1.68	–1.59	$P_2 \times P_5$	–9.54**	–9.00	$P_4 \times P_5$	2.39*	2.26
$P_1 \times P_4$	–0.12	–0.11	$P_2 \times P_6$	2.13*	2.01	$P_4 \times P_6$	–1.44	–1.36
$P_1 \times P_5$	–1.49	–1.41	$P_2 \times P_7$	1.47	1.39	$P_4 \times P_7$	–2.78*	–2.62
$P_1 \times P_6$	–0.38	–0.35	$P_3 \times P_4$	–1.49	–1.40	$P_5 \times P_6$	4.03**	3.81
$P_1 \times P_7$	2.60*	2.45	$P_3 \times P_5$	–0.10	–0.10	$P_5 \times P_7$	4.71**	4.45
$P_2 \times P_3$	1.44	1.36	$P_3 \times P_6$	1.74	1.64	$P_6 \times P_7$	–6.09**	–5.75

SE for sca effect of hybrids (*sij*) = $\sqrt{\dfrac{(P-3)}{(P-1)}Me'} = \sqrt{\dfrac{4}{6}\times 1.67} = 1.06$

SE $(s_{ij}-s_{ik}) = \sqrt{\dfrac{2(P-3)}{(P-2)}Me'} = \sqrt{\dfrac{8}{5}\times 1.67} = 1.65$

SE $(s_{ij}-s_{kl}) = \sqrt{\dfrac{2(P-4)}{(P-2)}Me'} = \sqrt{\dfrac{6}{5}\times 1.67} = 1.83$

t cal. = $\dfrac{gi}{SEgi}$ 't' table value at error df (40) is 2.02 at 5 % and 2.70 at 1.% level of significance.

CONCLUSION

The analysis of variance for combining ability revealed that variances due to gca and sca were highly significant for seed yield of chickpea. This indicated that both additive and non additive gene actions were involved in the inheritance of this trait. However, the ratio of gca /sca variance indicated that non additive gene action played major role in the inheritance of seed yield.

The estimates of *gca* effects revealed that parents viz., P_5, P_6 and P_7 had positive and significant *gca* effects for seed yield and hence found to be the best general combiners.

Five crosses showed significant high sca effects in desirable direction for grain yield. The cross combination viz., $P_5 \times P_6$ and $P_5 \times P_7$ with positive and significant *sca* effects for grain yield involved both the parents with high gca effects. Thus additive genetic variance was predominant and hence simple selection by pedigree method will be effective for further genetic improvement or for development of pure line variety from this cross.

High *sca* for grain yield was also recorded in the crosses *viz.*, $P_1 \times P_7$, $P_2 \times P_7$ and $P_4 \times P_5$ involving one parent with high and other with low *gca* effect indicating presence of non additive gene action. Therefore, recurrent selection method of breeding could be followed.

25 Partial Diallel Analysis

Partial diallel technique is developed by Kempthorne (1957) & further elaborated by Kempthorne and Curnow (1961). Diallel cross becomes unmanageable as the number of parents increases and if only a small number of inbreds are tested, the estimates of combining ability tend to have large sampling error. These difficulties have led to the development of the concept of sampling of the crosses produced by large number of inbreds without affecting the efficiency of the diallel technique. Partial diallel is a modified form of diallel design, where a large number of parents (up to 20) are involved but, only a sample of all possible crosses among them is used for analysis. The total number of crosses for analysis is $(p \times s)/2$, where, p= No. of parents, s= sampled crosses per array

Sampling procedure

For sampling , a constant 'K' (selection factor) is defined as $k =$

This constant should be whole number. For this both p and s can neither be odd or even number. If p is an even, s should be an odd number and vice versa. Further, s should be whole number (p/2)

Suppose,

Let us consider p=10 and s=5, then each parent is to be involved in 5 crosses. The total number of sampled crosses will be $10 \times 5/2$

k =(10+1–5)/2=3. Thus, the sampling is to begin after 3 arrays *i.e.* from the 4th array

Parent 1 × k +1, k +2,........, K+s : 1× 3+1, 1× 3+2, 1× 3+3, 1× 3+4, 1 × 3+5

i.e. 1×4,1×5,1×6,1×7,1×8

Parent p × k +p, k +p+1,........, K+p–1+s

Procedure for sampling and plan of crossing in partial diallel design: (25 crosses have to be made among 10 parents) is as follow.

Parents	P_1	P_2	P_3	P_4	P_5	P_6	P_7	P_8	P_9	P_{10}
P_1				X	X	X	X	X		
P_2					X	X	X	X	X	
P_3						X	X	X	X	X
P_4							X	X	X	X
P_5								X	X	X
P_6									X	X
P_7										X
P_8										
P_9										
P_{10}										

Genetic information:

Partial diallel provides information about *gca* and *sca* variances, *gca* effects, additive and dominance components but not *sca* effects.

Steps of partial diallel analysis

1. **Selection of parents**: Large number of parents (up to 20) are selected fro germplasm (inbreds, varieties, or newly developed cultures).
2. **Effecting of sampled crosses** : Crosses are made as per procedure given as above.
3. **Evaluation of crosses**: The sample crosses are evaluated in a standard statistical design (RBD) with required number of replications.
4. **Statistical analysis**: The mean data of biometrical characters recorded on all the genotypes are subjected to analysis of variance (RBD). The significance of F value for genotypes indicate significant differences among the genotypes studied and may be proceeded for combining ability analysis.
5. **Combining ability analysis**:

   (*i*) **Preparation of symmetrical circulant matrix: A** {Put diagonal elements equal to 's', sample crosses equal to 1 and other elements equal to 0)

$$\begin{bmatrix} 5 & 0 & 0 & 1 & 1 & 1 & 1 & 1 & 0 & 0 \\ 0 & 5 & 0 & 0 & 1 & 1 & 1 & 1 & 1 & 0 \\ 0 & 0 & 5 & 0 & 0 & 1 & 1 & 1 & 1 & 1 \\ 1 & 0 & 0 & 5 & 0 & 0 & 1 & 1 & 1 & 1 \\ 1 & 1 & 0 & 0 & 5 & 0 & 0 & 1 & 1 & 1 \\ 1 & 1 & 1 & 0 & 0 & 5 & 0 & 0 & 1 & 1 \\ 1 & 1 & 1 & 1 & 0 & 0 & 5 & 0 & 0 & 1 \\ 1 & 1 & 1 & 1 & 1 & 0 & 0 & 5 & 0 & 0 \\ 0 & 1 & 1 & 1 & 1 & 1 & 0 & 0 & 5 & 0 \\ 0 & 0 & 1 & 1 & 1 & 1 & 1 & 0 & 0 & 5 \end{bmatrix}$$

Its inverse is calculated as A^{-1}

(*ii*) Compute sum of corrected mean of sampled crosses (Qi)

(*a*) Correction factor = $\frac{2 \times GM}{P}$

Where, Gm = Grand mean of all the sampled crosses over replications

(*b*) The corrected mean of sampled cross:

= Uncorrected mean over replication– correction factor

(*c*) The corrected means of all sampled crosses in which a particular parent is involved are

pooled to obtain Qi values

Example :

Parent	Crosses with parent	Sum of their corrected mean (Qi)
P_1	$P_4, P_5, P_6, P_7, P_8,$	$(P_1 \times P_4) + P_1 \times P_5) + P_1 \times P_6) + P_1 \times P_7) + P_1 \times P_8)$

Sum of Qi= 0

(*iii*) Estimation of *gca* effect:

GCA effect (Gi) in matrix notaion is = AG= Q

$G= A^{-1}Q$

Where,

A^{-1} – Inverse matrix of A

Q – column vector of sum of corrected means of sampled crosses

G – column vector of gca effects

(*iv*) Analysis of variance for combining ability

Sum of squares due to GCA = $r\, G_iQ_i$ where, r= No. of replications

Sum of squares due to sca = Genotype SS– SS due to GCA

Analysis of variance for combining ability

Source	df	MS	Sum of squares
GCA	P–1	Mg	$\sigma_e^2 + r\sigma_S^2 + \frac{rs(p-2)}{(p-1)}\sigma_g^2$
SCA	$P\left[\frac{S}{2}-1\right]$	Ms	$\sigma_e^2 + r\sigma_S^2$
Error	$(r-1)\left(\frac{ps}{2}-1\right)$	Me	σ_e^2

(*v*) Genetic components

Variance due to GCA = Cov. HS

Variance due to SCA = Cov. FS – 2Cov. HS

$$\text{Variance due to GCA} = \sigma_g^2 = \frac{\text{Mg} - \text{Ms}}{\text{rs(p–2)/(p–1)}}$$

$$\text{Variance due to SCA} = \sigma_S^2 = \frac{\text{Ms} - \text{Me}}{r}$$

(*vi*) When the parents are inbreds or pure lines (F=1) and assuming no epistsis

$$\text{Variance due to GCA} = \text{Cov. HS} = \frac{1}{2}\sigma_A^2$$

$\sigma^2_A = 2\sigma^2_g$

Variance due to SCA = Cov FS – 2Cov. HS = σ^2_D

$\sigma^2_D = \sigma^2_S$

The ratio of σ^2_A/σ^2_D indicates the relative significance of additive versus non–additive genetic variance.

Significance of gca effects by t test

$$\text{Average variance (gi)} = \left[\frac{pa^0}{p-1} - \frac{1}{2s(p-1)}\right]\left[\frac{6_S^2 + 6_e^2}{r}\right]$$

A° is the diagonal element of A^{-1} =

SE gi = $\sqrt{\text{Average variance}}$

$$\text{Calculated 't'} = \frac{\text{Effect}}{\text{SE}}$$

SOLVED EXERCISE

A partial diallel was carried out with 8 parents in pearl millet. The sample size is 3. The grain yield data of 12 crosses obtained in RBD with four replication is analyzed as follows.

Data on Grain yield per plant (g)

S.N	Crosses	R–I	R–II	R–III	R–III	Total	Mean
1	$P_1 \times P_4$	128.10	123.84	92.56	115.28	459.78	114.945
2	$P_1 \times P_5$	128.36	119.84	103.24	129.72	481.16	120.290
3	$P_1 \times P_6$	74.40	70.86	60.94	68.00	274.20	68.550
4	$P_2 \times P_5$	91.52	113.96	87.26	106.98	399.72	99.930
5	$P_2 \times P_6$	59.06	65.62	81.62	86.76	293.06	73.265
6	$P_2 \times P_7$	84.16	109.74	102.14	94.52	390.56	97.640
7	$P_3 \times P_6$	109.86	98.16	93.26	102.26	403.54	100.885
8	$P_3 \times P_7$	117.20	100.28	116.16	112.52	446.16	111.540
9	$P_3 \times P_8$	109.68	116.48	123.92	120.86	470.94	117.735
10	$P_4 \times P_7$	53.40	60.86	74.46	69.08	257.80	64.450
11	$P_4 \times P_8$	53.86	48.30	40.64	44.62	187.42	46.855
12	$P_5 \times P_8$	86.62	94.18	90.32	108.16	379.28	94.820
	Total	1096.22	1122.12	1066.52	1158.76	4443.62	1110.91

Analysis of variance for RBD

Source	Df	SS	MS	F ratio
Replicates	3	383.46	127.82	1.27
Genotypes	11	24985.97	2271.45**	22.56
Error	33	3221.57	100.65	
Total	47	28691.00		

There is significant differences among the sampled crosses for grain yield, hence proceeded for combining ability analysis.

Estimation of combining ability effects.

Parents	crosses	Sum of mean of crosses (S_1)	CF	$Qi=S_1-CF$
1	4, 5, 6	303.79	277.73	26.06
2	5, 6, 7	270.84	277.73	–6.90
3	6, 7, 8	330.16	277.73	52.43
4	7, 8, 1	226.25	277.73	–51.48
5	8, 1, 2	315.04	277.73	37.31
6	1, 2, 3	242.70	277.73	–35.03
7	2, 3, 4	273.63	277.73	–4.10
8	3, 4, 5	259.41	277.73	–18.32

(*a*) Correction factor = $\frac{2 \times GM}{P} = \frac{2 \times 1110.91}{8} = 277.73$

Estimation of *gca* effect:

gca effect (G*i*) in matrix notation is = AG= Q

$G = A^{-1}Q$

After solving the above matrices the GCA effects of parents are

Parent	gca effect		Parent	gca effect
P_1	24.60		P_5	5.21
P_2	13.16		P_6	–37.60
P_3	40.03		P_7	–13.98
P_4	–15.34		P_8	–16.07

Combining Ability Analysis

Sum of squares due to GCA = $r\,G_iQ_i$ = 21208.97

Sum of squares due to SCA = Genotype SS – SS due to GCA = 24985.96–21208.97

= 3776.99

Analysis of Variance for Combining Ability

Source	Df	SS	MS	F
GCA	7	21208.97	3029.85** (Mg)	30.10
SCA	4	3776.99	944.24** (Ms)	9.38
Error	33	3221.57	100.63 (Me)	

$$\sigma^2 GCA = \frac{Mg - Ms}{rs(p-2)/(p-1)} = \frac{3029.85 - 944.24}{4 \times 3 \times 6/(7)} = 202.78$$

$$\text{Variance due to SCA} = \sigma^2_5 = \frac{Ms - Me}{r} = 210.90$$

$\sigma^2_D = 2\sigma^2_g = 2 \times 202.78 = 405.56$

$\sigma^2_D = \sigma^2_S = 210.90$

$\sigma^2_A = \frac{\sigma^2_A}{\sigma^2_D}\,\sigma^2_D = 1.92$

The ratio of σ^2_A/σ^2_D is greater than one which indicates that the grain yield in ppearl millet is governed by additive gene action. Therefore, simple selection techniques and recombination breeding with pedigree method of selection is suggested for grain improvement.

Significance of gca effects

$$\text{Average variance (gi)} = \left[\frac{pa^0}{p-1} - \frac{1}{2s(p-1)}\right]\left[\frac{6^2_S + 6^2_e}{r}\right]$$

A^o is the diagonal element of A^{-1} =

$$= \left[\frac{8 \times 0.94}{7} - \frac{1}{2 \times 3 \times 7}\right]\left[\frac{944.24 + 100.65}{4}\right] = 274.41$$

$T = \frac{gi}{SE}$ calculated t at 33 error df =2.03

Crosses with parent	Parent Mean	Qi	GCA
P_1	101.26	26.055	24.60
P_2	90.28	–6.895	13.16
P_3	110.05	52.43	40.03*
P_4	75.42	–51.48	–15.34
P_5	105.01	37.31	5.21
P_6	80.90	–35.03	–37.60*
P_7	91.21	–4.1	–13.98
P_8	86.47	–18.32	–16.07

Parents P_1, P_2 and P_3 recorded positive effects, however only parent P3 had significant effect. The parents P_1 and P_3 may be used in recombination breeding with pedigree method of selection.

26 Triallel and Quadriallel Analysis

Rawlings and Cockerham (1962 a & b) developed the concept of triallel and quadriallel. These techniques are useful for obtaining information on components of epistasis variance in addition to additive and dominance variance. A description of these techniques is as below.

TRIALLEL ANALYSIS

This also refers to analysis of a three–way cross. Triallel analysis includes all possible three–way crosses among *n* parents. A three way cross symbolized by (A × B) × C and defined as a cross between the line C and unrelated F_1 hybrid (A × B). Lines A and B are grand parental lines and line C is full or immediate parental line. Main features of triallel analysis are

(*i*) Triallel analysis includes all possible three–way crosses among *n* parents. In triallel analysis, total number of three way crosses is equal to

$$\frac{n(n-1)(n-2)}{2}$$

(*ii*) Each cross involves three different parents.

(*iii*) It provides information about additive, dominance and epistasis components.

(*iv*) It measures general and specific line effects and helps in deciding mating order of parents in three –way cross combinations.

Steps in Analysis of triallel cross

1. Effecting single crosses: Few numbers of inbred lines are selected as parents. In first crop season selected parents are mated in a half diallel fashion to obtain single crosses. With five parents there will be 10 single crosses.
2. Effecting three way crosses: All single crosses mated with the parents to obtain three way crosses in such a way that no parent should be appeared more than once in the same three way cross. With five parents there will be 30 three way crosses.
3. Evaluation of crosses: All three way crosses are evaluated in RBD with required number of replication. Replication wise biometrical observations recorded are used for statistical analysis.
4. Statistical analysis; Analysis of triallel cross data is carried out according to the procedure given by Singh and Chaudhary (1985).

Genetic information obtained

Triallel analysis provides information about combining ability effects and components of genetic variance.

Combining ability effects: Two types of combining ability effects are estimated *viz.*, general line effect and specific line effect.

General line effect: General line effects are two types

(*i*) General line effect of first kind (hi) which refers to the general combining ability effect of a line used as one of the grand parent.

(*ii*) General combining ablity effect of second kind (gi) which refers to the general combining ability effect of a line used as a parent crossed to the single cross hybrid.

Specific line effects : There are three types of specific line effects as below

(*i*) Two line specific effect of first kind (dij) : It refers to the specific combining ability effect of a line used as one of the grand parent.

(*ii*) Two line specific effect of second kind (sik) : It refers to the specific combining ability effect of a line when crossed as a parent to the single cross hybrid.

(*iii*) Three line specific effect (tijk) : It refers to the specific combining ability effect of a line in three way crosses.

Plan of crossing for triallel analysis

Single crosses	**Male parents**				
	P_1	P_2	P_3	P_4	P_5
$P_1 \times P_2$	–	–	$(P_1 \times P_2) P_3$	$(P_1 \times P_2) P_4$	$(P_1 \times P_2) P_5$
$P_1 \times P_3$	–	$(P_1 \times P_3) P_2$	–	$(P_1 \times P_3) P_4$	$(P_1 \times P_3) P_5$
$P_1 \times P_4$	–	$(P_1 \times P_4) P_2$	$(P_1 \times P_4) P_3$	–	$(P_1 \times P_4) P_5$
$P_1 \times P_5$	–	$(P_1 \times P_5) P_2$	$(P_1 \times P_5) P_3$	$(P_1 \times P_5) P_4$	–
$P_2 \times P_3$	$(P_2 \times P_3) P_1$	–	–	$(P_2 \times P_3) P_4$	$(P_2 \times P_3) P_5$
$P_2 \times P_4$	$(P_2 \times P_4) P_1$	–	$(P_2 \times P_4) P_3$	–	$(P_2 \times P_4) P_5$
$P_2 \times P_5$	$(P_2 \times P_5) P_1$	–	$(P_2 \times P_5) P_3$	$(P_2 \times P_5) P_4$	–
$P_3 \times P_4$	$(P_3 \times P_4) P_1$	$(P_3 \times P_4) P_2$	–	–	$(P_3 x P_4) P_5$
$P_3 \times P_5$	$(P_3 \times P_5) P_1$	$(P_3 \times P_5) P_2$	–	$(P_3 \times P_5) P_4$	–
$P_4 \times P_5$	$(P_4 \times P_5) P_1$	$(P_4 \times P_5) P_2$	$(P_4 \times P_5) P_3$	–	–

Triallel cross requires two crop seasons for generating experimental material only and also more crosses have to be made than in diallel cross which adds to the experimental cross.

The procedure of combining ability for triallel cross and estimation of genetic components is given by Singh and Chaudhary (1985).

27 Quadriallel Analysis

This also refers to analysis of double cross hybrids. Quadriallel analysis includes all possible double crosses among n parents. A double cross symbolized by (A × B) × (C × D) and defined as a cross between two unrelated F_1 hybrids. Parents are A, B, C and D and and (A × B and C × D) are two unrelated F_1 hybrids. Main features of quadriallel analysis are

(*i*) Quadriallel analysis includes all possible double crosses among n parents. In quadriallel analysis, total number of double crosses is equal to

$$\frac{n(n-1)(n-2)(n-3)}{8}$$

(*ii*) Each cross involves four different parents.

(*iii*) It also provides information about additive, dominance and epistasis components.

(*iv*) It measures one, two, three and four lines average effects and helps in deciding mating order of parents in double cross hybrids in cross pollinated crops

Steps in Analysis of Quadriallel Cross

1. Effecting single crosses: Few number of inbred lines are selected as parents. In first crop season selected parents are mated in a half diallel fashion to obtain single crosses. The number of crosses among n parents would be equal to n(n–1)/2. With five parents there will be 10 single crosses among them
2. Effecting double crosses: The F_1 crosses are again mated in diallel fashion to produce double cross hybrids with the restriction that no parent should be appeared more than once in the same double cross combination. With five parents there will be 15 double crosses.
3. Evaluation of crosses: All double crosses are evaluated in RBD with required number of replication. Replication wise biometrical observation recorded are used for statistical analysis .
4. Statistical analysis; Analysis of quadriallel cross data is carried out according to the procedure given by Singh and chaudhary (1985).

Genetic Information Obtained:

Quadriallel analysis provides information about combining ability effects and components of genetic variance.

Combining ability effects: The following types of combining ability effects are estimated.

(*i*) Average effect of line I

(*ii*) Two interaction effect of lines i and j appearing together irrespective of arrangement

(*iii*) Three–line interaction effects of lines i,j and k appearing together irrespective of arrangement.

(*iv*) Four –line interaction effect of lines i,j,,k and i appearing together irrespective of arrangement.

(*v*) 2–line interaction effect of lines i and j due to particular arrangement (ij) (– –)

(*vi*) 2–line interaction effect of lines i and j due to particular arrangement (i–)(j–)

(*vii*) 3–line interaction effect of lines I,j and k due to particular arrangement, (ij) (k–)

(*viii*) 4–line interaction effect of lines I,j,k and l due to particular arrangement (ij) (kl)

Plan of crossing (Double cross hybrids)

	$(P_2 \times P_3)$	$(P_2 \times P_4)$	$(P_2 \times P_5)$	$(P_3 \times P_4)$	$(P_3 \times P_5)$	$(P_4 \times P_5)$
$(P_1 \times P_2)$	–	–	–	X	X	X
$(P_1 \times P_3)$	–	X	X	–	–	X
$(P_1 \times P_4)$	X	–	X	–	X	–
$(P_1 \times P_5)$	X	X	–	X	–	–
$(P_2 \times P_3)$	–	–	–	–	–	X
$(P_2 \times P_4)$	–	–	–	–	–	X
$(P_2 \times P_5)$	–	–	–	X	–	–

The analysis of double–cross hybrids, *i.e.*, quadriallel analysis required two cropping seasons and also more labour as comparatively large numbers of crosses are to be made in contrast to diallel and triallel analysis.

The procedure of combining ability for double cross and estimation of genetic components is given by Singh and Chaudhary (1985).

28 Biparental Cross Analysis

In mating designs *viz.*, top cross, poly cross, diallel, triallel, quadriallele, partial diallel and L × T, the parents selected for crossing are generally pure lines or inbreds (Homozygous parents). However, in random mating designs, the parents are not pure lines but heterozygous segregating individuals (randomly selected in F_2 or subsequent generation of a cross between two pure lines having contrasting performance). The selected heterozygous segregating individual plants are crossed in different definite fashion. Depending upon the method of crossing between individual plants, there are two matting designs *viz.*, (1) Biparental cross and (ii) Triple test cross.

I. BIPARENTAL CROSS

The concept of biparental mating was originally proposed by Comstock and Robinson (1948,1952). In biparental mating, large F_2 population of a cross is grown and several single plants are randomly selected for crossing. Based on the fashion of crossing between selected F_2 plants, there are three mating designs of biparental cross. These methods are known as North Carolina Design, I, II and III. In all these designs, full sib and half sib progenies are produced by attempting biparental cross in F_2 generation of a cross.

Utility of designs

1. Each of these designs provides estimates of additive genetic variance and dominance genetic variance.
2. Biparental mating is generally used in several crop plants mainly for population improvement.
3. It has been successful used for creating variability and breaking some undesirable linkages in many crops.

A. North Carolina Design I

It is a very popular multipurpose design for both theoretical and practical plant breeding applications. It is commonly used to estimate additive and dominance variances as well as for the evaluation of full- and half-sib recurrent selection. It requires sufficient seed for replicated evaluation trials, and hence is not of practical application in breeding species that are not capable of producing large amounts of seed. It is applicable to both self- and cross-pollinated species that meet this criterion. It has been extensively used in maize breeding for estimating genetic variances.

Mating procedure

From the F_2 population, a individual randomly selected and used as a male. A set of 4 randomly selected plants are used as females and are mated to the above male. Thus, set of 4 full sib families are produced. This is denoted as male group. Similarly large numbers of male groups are produced. No female is used for second mating. Four male groups (16 female groups) form a set. Several such sets are made for evaluation. Each set form an experimental unit i.e block with m × f plots in a replication.

Total crosses available for evaluation= No. of females used in a set (mf) × No. of set.

Evaluation and statistical analysis of crosses.

The crosses are evaluated in RBD with required number of replications. The mean data of a characters recorded on the single plants in each cross are subjected to appropriate analysis of variance (RBD). After significance of F test, combining ability analysis is to be carried out.

Combining Ability Analysis: The analysis of variance for combining ability along with expectation of mean squares are as below

The ANOVA for NCD I

Source	df	MS	Expectation of MS
Sets	(*s*-1)		
Replication in sets	*s*(*r*-1)		
Males in sets	*s*(*m*-1)	MS_1	$\sigma^2_e + r\sigma^2_f + rf\sigma^2_m$
Females in males in sets	*sm*(*f*-1)	MS_2	$\sigma^2_e + r\sigma^2_f$
Error	s(*mf*-1) (*r*-1)	MS_3	σ^2_e
Total	(smfr-1)		

Genetic components

In this design, the variance among single crosses is divided into two fractions. Since the parents are non inbreds the inbreeding coefficient F= o. Assuming absence of epistasis interaction,

(*i*) Variance due to male (Cov. HS) $= \acute{\sigma}^2 m = \frac{1}{4}\sigma^{2A}$

(*ii*) Variance due to females in male (Cov FS- Cov. HS) $= \frac{1}{2}\sigma^2_A + \frac{1}{4}\sigma^2_D - \frac{1}{4}\sigma^2_A$

$= \frac{1}{4}\sigma^2_A + \frac{1}{4}\sigma^2_D$

The different genetic components are derived as below

$\hat{\sigma}^2 m = \frac{MS_1 - MS_2}{r \times f}$ = Cov. HS; $\hat{\sigma}^2 m = \frac{1}{4\sigma^2_A}$

$\sigma^2_A = 4\hat{\sigma}^2 m$

$\hat{\sigma}^2 f = \frac{MS_2 - MS_3}{r}$ = Cov. FS – Cov HS; $\hat{\sigma}^2 f = \frac{1}{4}\sigma^2_A + \frac{1}{4}\sigma^2_D$

$\sigma^2_D = 4\hat{\sigma}^2 f - 4\hat{\sigma}^2 m$

The ratio of / suggests the relative importance of additive versus non additive genetic variance.

SOLVED EXERCISE NCD I

In a F_2 population of maize, 24 full sib progenies were produced by crossing 12 random males (3 sets with 4 males in each set) each with 2 different random females. These 24 progenies were evaluated in RBD with 2 replications. The mean data given on grain yield of progenies of each full sib family given below is analyzed for estimation of genetic components as per procedure of NCD-I.

Data on Grain yield/plan

		RI		RII		Total		
Set	Male	Female-1	Female-2	Female-1	Female-2	Female-1	Female-2	Total
I	1	32.4	30.5	28.5	27.4	60.9	57.9	118.8
	2	44.3	32.2	46.9	28.5	91.2	60.7	151.9
	3	24.6	21	27.2	22.8	51.8	43.8	95.6
	4	44.3	35.6	41.9	30.1	86.2	65.7	151.9
	Total		264.9		253.3			518.2
II	1	36.9	34.6	33.5	31.2	70.4	65.8	136.2
	2	47.7	43.2	43.2	40.3	90.9	83.5	174.4
	3	45.4	46.3	46	49.2	91.4	95.5	186.9
	4	52.6	51.2	50.6	48.9	103.2	100.1	203.3
	Total		357.9		342.9			700.8
III	1	29.4	38.6	32.2	35.4	61.6	74	135.6
	2	25.8	25.2	21.6	23.6	47.4	48.8	96.2
	3	31.8	34.8	33.2	37.4	65	72.2	137.2
	4	46.3	42.6	44.2	37.3	90.2	79.9	170.1
	Total		274.5		264.9			539.1
							GT	1758.1

Estimation of sum of squares due to different sources of variation

$$CF = \frac{(GF)^2}{S \times r \times m \times} = \frac{(1758.1)^2}{48} = 64394.08$$

Total SS = $32.4^2 + 30.5^2 + ... + 37.3^2$ – CF = 68114.28- 64394.08 =3720.20

$$\text{Set SS} = \frac{(518.2)^2 + (700.8)^2 (539.1)^2}{r \times m \times f} - \text{CF} = \frac{1050280.69}{16}$$

$= 65642.54 - 64394.08 = 1248.46$

$$\text{Rep in Set SS} = \frac{(264.9)^2 + ... + (264.9)^2}{m \times f} - \text{CF} - \text{Set SS} = \frac{525527.98}{8} - \text{CF} - \text{Set SS}$$

$= 65690.99 - 64394.08 - 1248.46 = 48.46$

$$\text{Males in Set SS} = \frac{(118.8)^2 + ... (170.1)^2}{r \times f} - \text{CF} - \text{Set SS} = \frac{270027.97}{4} - \text{CF} - \text{Set SS}$$

$= 67506.99 - 64394.08 - 1248.46 = 1864.45$

$$\text{Females in males Set SS} = \frac{(60.9)^2 + (57.9)^2 + ... + (79.9)^2}{r} - \text{CF} - \text{Males in sets SS} - \text{Set SS}$$

$= 67966.87 - 64394.08 - 1864.45 - 1248.46 = 459.88$

Error SS = Total SS - Females in males in sets SS – Males in sets SS – Repl. In set SS – Sets SS

$= 3720.20 - 459.88 - 1864.45 - 48.46 - 1248.46 = 98.95$

ANOVA for NCD I

Source	df		SS	MSS		F cal.
Sets	(s-1)	2	1248.46	624.23**		132.53
Rep. in sets	s(r-1)	3	48.46	16.15**		3.43
Males in sets	s(m-1)	9	1864.45	207.16**	MS_1	43.98
Females in males in sets	sm(f-1)	12	459.88	38.32**	MS_2	8.14
Error	s(mf-1)(r-1)	21	98.95	4.71	MS_3	
Total	(smfr-1)	47	3720.20			

Genetic parameters

$$\sigma^2 m = \frac{MS_1 - MS_2}{f \times r} = \frac{207.16 - 38.32}{2 \times 2} = 42.21$$

$$\sigma^2 f = \frac{MS_2 - MS_3}{r} = \frac{38.32 - 4.71}{2} = 16.81$$

$\sigma^2\text{m} = \frac{1}{4}\, \sigma^2\text{A}$

hence, σ^2A (Additive genetic variance) = $4 \times \sigma^2\text{m} = 168.84$

σ^2D (Dominance genetic variance) $= 4 \times \sigma^2\text{f} - \sigma^2_A$

$= (4 \times 16.81) - 168.84 = -101.60$

CONCLUSION

The additive variance is more than dominance variance, grain yield is governed by the additive gene action. The predominance of additive gene action suggests that progeny selection could be used effectively to improve the population for grain yield. Moreover, an approach of combining the favourable additive gene present in the full-sib progenies into a single pool by developing composite or synthetics is suggested for improvement in Maize.

B. North Carolina Design II

Mating procedure

In this design, from base population of F_2, *m* males and *f* females are randomly selected and each male is crossed each of the females. Thus, $m \times f$ crosses will form a set. Many such sets can be made for evaluation. However, for each set, male and female plants are selected afresh. The design is most adapted to plants that have multiple flowers. The design is essentially a two-way ANOVA in which the variation may be partitioned into difference between males (*m*) and females (*f*) and their interaction.

The ANOVA NCD II

Source	df	MS	Expectation of MS
Sets	$(s-1)$		
Replication in sets	$s(r-1)$		
Between males in sets	$s(m-1)$	MS_1	$\sigma^2_e + r\,\sigma^2_{mf} + rf\sigma^2_m$
Between females in sets	$s(f-1)$	MS_2	$\sigma^2_e + r\,\sigma^2_{mf} + rmf\sigma^2_f$
Males x females in sets	$s(m-1)(f-1)$	MS_3	$\sigma^2_e + r\,\sigma^2_{mf}$
Error	$s(mf-1)(r-1)$	MS_4	σ^2_e
Total	$(smfr-1)$		

Genetic components

Since the parents are non inbreds the inbreeding coefficient F= 0. Assuming absence of epistasis interaction,

Variance due to male (Cov. HS) = $\frac{1}{4}\sigma^2_A$

Variance due to female (Cov. HS) = $\frac{1}{4}\sigma^2_A$

$$\text{Average Cov. HS} = \frac{(MS_1 - MS_3) + (MS_2 - MS_3)}{r(m+f)} = \frac{1}{4}\sigma^2_A$$

Variance due to males x females = (Cov FS- 2Cov. HS) = σ^2_{mf}

$$\frac{(MS_3 - (MS_4)}{r} = \frac{1}{4}\sigma^2_D \text{ hence, } \sigma^2_D = 4\,\sigma^2 mf$$

The ratio of σ^2_A/σ^2_D suggests the relative importance of additive versus non additive genetic variance.

SOLVED EXERCISE NCD II

In a F_2 population of bajra, 40 full sib progenies were produced by crossing 5 random males each with 4 same random females (Two sets) . These 40 progenies were evaluated in RBD with 2 replications. The mean data on grain yield of progenies of each full sib family is recorded and analysis of combining analysis (Genetic components) is carried out as per NCD II

Set I

	RI				RII				Total				
Male	F1	F2	F3	F4	F1	F2	F3	F4	F1	F2	F3	F4	Total
1	32.1	23.2	35.6	22.3	34.3	21.2	36.4	24.1	66.4	44.4	72.0	46.4	229.2
2	19.6	35.6	22.4	28.6	21.6	37.6	24.2	26.6	41.2	73.2	46.6	55.2	216.2
3	35.2	31.6	34.6	34.9	37.2	29.6	35.8	33.2	72.4	61.2	70.4	68.1	272.1
4	49.4	42.1	45.8	44.6	51.6	44.1	43.8	42.4	101.0	86.2	89.6	87.0	363.8
5	43.2	48.5	50.2	40.3	45.3	47.3	48.2	43.2	88.5	95.8	98.4	83.5	366.2
T				719.8				727.7	369.5	360.8	377.0	340.2	1447.5

Set-II

	RI				RII				Total				
Male	F1	F2	F3	F4	F1	**F2**	**F3**	**F4**	**F1**	**F2**	**F3**	**F4**	**Total**
1	20.2	35.4	32.2	24.3	24.3	33.2	32.0	23.2	44.5	68.6	64.2	47.5	224.8
2	42.6	38.6	26.4	30.2	40.6	35.6	21.4	29.4	83.2	74.2	47.8	59.6	264.8
3	38.3	40.3	30.6	36.7	37.2	33.2	28.5	35.5	75.5	73.5	59.1	72.2	280.3
4	48.5	45.1	30.2	42.5	51.6	44.2	32.1	42.2	100.1	89.3	62.3	84.7	336.4
5	40.2	50.3	46.5	43.2	45.3	49.3	45.6	41.6	85.5	99.6	92.1	84.8	362.0
T				742.3				726.0	388.8	405.2	325.5	348.8	1468.3
												Total	2915.80

Estimates of SS due to different sources

$$CF = \frac{(GT)^2}{S \times r \times m \times f} = \frac{(2915.80)^2}{80} = 106273.62$$

$$\text{Total SS} = 32.1^2 + 23.2^2 + \dots + 41.6^2 - CF = 6120.60$$

$$\text{Set SS} = \frac{(1447.5)^2 + (1468.3)^2}{r \times m \times f} - CF = 5.41$$

$$\text{Rep in Set SS} = \frac{(719.8)^2 + \dots (726)^2}{m \times f} - CF - \text{Set SS} = 8.20$$

$$\text{Males in Set SS} = \frac{(229.2)^2 + \ldots + (348.0)^2}{r \times m} - \text{CF} - \text{Set SS} = 4120.66$$

$$\text{Females in Set SS} = \frac{(369.2)^2 + \ldots (348.8)^2}{r \times m} - \text{CF} - \text{Set SS} = 474.58$$

$$\text{Males x Females in Set SS} = \frac{(66.4)^2 + \ldots + (84.8)^2}{\text{r}} - \text{CF} - \text{Set SS} - \text{male SS} - \text{Female SS}$$

$$= 1404.08$$

Error SS = TSS - SetSS - Rep SS - Male SS - Female SS - M x F SS=111.67

ANOVA

Sources	df		SS	MS		F cal
Rep in stes	s(r-1)	2	8.20	4.10		0.87
Sets	(s-1)	1	5.41	5.41		1.15
Males in sets	s(m-1)	8	4120.66	515.08	MS_1	109.36
Female s in sets	s(f-1)	6	474.58	79.10	MS_2	16.79
M × F in sets	s(m-1)(f-1)	24	1404.08	58.50	MS_3	12.42
Error	s(mf-1)(r-1)	38	111.67	2.94	MS_4	
Total	(smfr-1)	79	6124.60			

Genetic parameters

$$\sigma^2 m = \frac{MS_1 - MS_3}{r \times f} = \frac{515.08 - 58.5}{2 \times 4} = 57.07$$

$$\sigma^2 f = \frac{MS_2 - MS_3}{r \times m} = \frac{79.10 - 58.50}{2 \times 5} = 2.06$$

$$\sigma^2 m\ x\ f\ in\ sets = \frac{MS_3 - MS_4}{\text{r}} = \frac{58.50 - 2.94}{2} = 27.78$$

$\sigma^2 m = ¼\, \sigma^2_A \,.\, \sigma^2 f = ¼\, \sigma^2_A$

hence, σ^2_A (Additive genetic variance) = 4(σ^2m+ σ^2f) /2 = 118.26

$\sigma^2 mf = ¼\, \sigma^2_D$

σ^2_D (Dominance genetic variance) = 4 × $\sigma^2 mf$ = 111.13

Ratio σ^2_A / σ^2_D = 1.06

CONCLUSION

As the ratio of σ^2_A / σ^2_D is more than 1 (1.06), grain yield is governed by the additive gene action. Hence, it is suggested that that progeny selection could be used effectively to improve the population for grain yield. More ever, an approach of combining the favourable additive gene present in the full-sib progenies into a single pool by developing composite or synthetics is suggested for improvement in bajra.

C. North Carolina Design III

In this design, randomly selected F_2 plants is crossed with the parental lines (P_1 and P_2) from which the F_2 was descended. Hence, each male group consists of two male progenies and a set contains only 2n full sib progenies. Many such sets can be made for evaluation. No. of crosses obtained for evaluation= 2ns, where, n is the number of male plants per set and s is the number of set.

Among all NCD designs, it is most powerful and it can be employed to know the effect of linkage on the estimate of average degree of dominance.

Females

P_1

F_2 plants (males) are mated with → P_1 / P_2

P_2

The ANOVA NCD III

Source	df	MS	Expectation of MS
Sets	(*s*-1)		
Replication in sets	*s*(*r*-1)		
Males in sets	*s*(*n*-1)	MS_1	$\sigma^2_e + r\,\sigma^2_m$
Males x females in sets	*s*(*n*-1)	MS_2	$\sigma^2_e + r\,\sigma^2_{ml}$
Error	*s*(2*n*-1) (*r*1)	MS_3	σ^2_e
Total	(2*snr*-1)		

Genetic components

Since the parents are non inbreds the inbreeding coefficient F= 0. Assuming absence of epistasis interaction,

Variance due to male (Cov. HS) = $\sigma^2 m = \frac{1}{4}\sigma^2_A$

Variance due to males x females = (Cov FS- 2Cov. HS) = $\sigma^2 ml = \frac{1}{4}\sigma^2_D$

$$\sigma^2 m = \frac{MS_1 - MS_3}{2r} = \frac{1}{4}\sigma^2_A \text{ hence, } \sigma^2_A = 4\,\sigma^2 m$$

$$\sigma^2 mf = \frac{MS_2 - MS_3}{r} = \frac{1}{2}\sigma^2_D \text{ hence, } \sigma^2_D = 2\,\sigma^2 ml$$

The ratio of σ^2_A / σ^2_D suggests the relative importance of additive versus non additive genetic variance.

SOLVED EXERCISE NCD III

An a F2 population is produced by crossing two inbred lines L_1 and L_2. From F2, 12 individual plants (males) were selected randomly (4 in each set) and back crossed to both of inbreds (females). These 24 full sib families as treated as three sets and were evaluated in RBD with 2 replications. The mean data on grain yield of progenies of each full sib family is given below is analyzed as per the method NCD II

		RI		RII		Total		
Set	**Male**	**Female 1**	**Female 2**	**Female 1**	**Female 2**	**Female 1**	**Female 2**	**Total**
I	1	22.40	20.50	24.10	21.50	46.50	42.00	88.50
	2	38.20	30.40	35.10	32.40	73.30	62.80	136.10
	3	40.90	35.80	38.20	33.40	79.10	69.20	148.30
	4	26.90	24.10	28.40	26.40	55.30	50.50	105.80
	Total		**239.20**		**239.50**	**254.20**	**224.50**	**478.70**
II	1	30.40	35.20	34.20	31.40	64.60	66.60	131.20
	2	45.20	44.60	45.10	42.40	90.30	87.00	177.30
	3	35.60	32.90	36.40	31.80	72.00	64.70	136.70
	4	48.40	46.20	50.20	47.30	98.60	93.50	192.10
	Total		**318.50**		**318.80**	**325.50**	**311.80**	**637.30**
III	1	25.40	22.30	27.30	24.30	52.70	46.60	99.30
	2	33.20	28.40	35.10	30.80	68.30	59.20	127.50
	3	45.20	41.30	42.60	43.20	87.80	84.50	172.30
	4	46.90	44.10	48.30	46.40	95.20	90.50	185.70
	Total		**286.80**		**298.00**	**304.00**	**280.80**	**584.80**
								1700.80

Estimates of SS due to different sources

$$\text{CF} = \frac{(GT)^2}{S \times r \times m \times f} = \frac{(1700.80)^2}{48} = 60265.01$$

Total SS = $22.4^2 + 20.5^2 + ... + 46.4^2$ – CF = 3433.17

$$\text{Set SS} = \frac{(478.70)^2 + (637.30)^2 + (584.8)^2}{r \times m \times f} - \text{CF} = 815.99$$

$$\text{Rep in Set SS} = \frac{(239.20)^2 + ... + (280.8)^2}{\text{m x f}} - \text{CF} - \text{Set SS} = 7.85$$

$$\text{Females in Set SS} = \frac{(239.20)^2 + ... + (280.8)^2}{m \times f} - \text{CF} - \text{Set SS} = 100.50$$

$$\text{Males in Set SS} = \frac{(88.5)^2 + ... + (185.7)^2}{r \times f} - \text{CF} - \text{Set SS} = 2335.64$$

$$\text{Males} \times \text{Females Int in Set SS} = \frac{(46.5)^2 + ... + (90.5)^2}{\text{r}} - \text{CF} - \text{Set SS} - \text{male SS-}$$

Female SS

= 124.68

Error SS = TSS – Set SS – Rep SS – Male SS – Female SS – M × F SS = 48.51

ANOVA for NCD III

Source	df		SS	MSS		F cal.
Sets	(s-1)	2	815.99	407.99**		86.62
Rep. in sets	s(r-1)	3	7.85	2.62		0.56
Females in sets	s(f-1)	3	100.50	33.50**		7.11
Males in in sets	s(m-1)	9	2335.64	259.52**	MS_1	55.10
M x f int. in sets	s (m-1)(f-1)	9	124.68	13.85*	MS_2	2.94
Error	s(mf-1)(r-1)	21	48.51	2.31	MS_3	
Total	(smfr-1)	47	3433.17			

Genetic components

$$\sigma^2 m = \frac{MS_1 + MS_3}{fr} = \frac{259.52 - 2.31}{2 \times 2} = 64.30$$

$$\sigma^2 mf = \frac{MS_2 - MS_3}{r} = \frac{13.85 - 2.31}{2} = 5.77$$

$$\sigma^2 m = \frac{1}{4}\sigma_A^2$$

hence, $= \sigma_A^2 = 4\sigma^2 m \times 64.30 = 257.21$

hence, $= 2\sigma^2 mf = 2 \times 5.77 = 11.54$

Ratio $\sigma_A^2/\sigma_D^2 = 22.28$

CONCLUSION

Since Ratio is more than 1 (22.28), grain yield is governed by the additive gene action. The predominance of additive gene action suggests that progeny selection could be used effectively to improve the population for grain yield. More ever, an approach of combining the favourable additive gene present in the full-sib progenies into a single pool by developing composite or synthetics is suggested for improvement in bajra.

2. TRIPLE TEST CROSS

This was first devised by Kearsy and Jinks (1968). It is the extension of NCD III design. In this design, selected F_2 individuals are crossed with two inbred parents and their F_1. Hence, there are 3n families, where, n is the number of F_2 plants used for crossing. The evaluation of crosses and statistical analysis is same as in NCD I

The analysis of this design may be divided into two parts, the first part supplies a test for epistasis, and the second assesses the significance, and provides estimates of the additive and dominance components of variation. It is reported that the main use of the triple test cross will be to investigate the inheritance of quantitative traits in natural population and relating this to natural selection and ecology however, in practice the triple test cross is viewed as a resource consuming design.

For the biometrical geneticist, interest in determining the genetic architecture of a particular character and in estimating the genetic component with maximum precision, the triple test cross is without doubt the most powerful of the available mating designs.

		Females	
	→	P_1	–L1
F2 plants (males) are mated with	→	P_2	– L2
	→	F_1	–L3

Test of Epistasis

Assumption:

$L_{1i} + L_{2i} - 2L_{3i} = 0$ If this is 0, there is no epistasis.

Where,

L_{1i} = Mean of family obtained by crossing no. of F_2 plants to P_1

L_{2i} = Mean of family obtained by crossing no. of F_2 plants to P_2

L_{3i} = Mean of family obtained by crossing no. of F_2 plants to F_1

Test of significance

$$F = \frac{\Sigma(L_{1i} + L_{2i} - 2L_{3i})^2}{n(VL_1 + VL_2 + VL_3)}$$

n= number of F_2 plants selected

VL_1 = Average variance of L_{1i}

VL_2 = Average variance of L_{2i}

VL_3 = Average variance of L_{3i}

DF for F test= 3n (r–1)

Estimation of genetic components

In triple test cross analysis three comparisons are possible

$\bar{L}_{1i} + \bar{L}_{2i} + \bar{L}_{3i}$ = A (Additive component)

$\bar{L}_{1i} - \bar{L}_{2i}$ = D (Dominance component)

$\bar{L}_{1i} + \bar{L}_{2i} - \bar{L}_{3i}$ = I (Epistatic interaction component)

Modification of triple test cross model:

In modified triple test cross, instead of three testers only one tester is used (Jinda *et al.*, 1976). Further, Ketata *et al.* (1976) proposed modified approach of triple test cross in which, testers L_1, L_2 and their hybrid L_3 are crossed with a number of unrelated strains, instead of crossing with randomly selected F_2 individuals of the cross L_1 x L_2. This modified scheme has similar statistical treatment and yield comparable information. For worked out example, please refer Singh and Chaudhary (1985).

29

Transformation of data

Scale or Arithmetic unit: Most convenient unit of measurement of the character is called as scale or arithmetic unit. e.g. cm, kg, ft, number of days etc. Proper choice of scale is important and choice should be made in relation to the objective.

Transformation: The conversion of original unit to some other function according to what is more appropriate for the purpose of data to be used is called as transformation.

Transformation of the scale is necessary to bring the data into normal distribution to make the variance independent of mean and to remove/reduce non-additive interaction. In other words, transformation is required for removal of scale effect.

Scale effect: When the transformation is required but not done that time certain phenomenon arise such as non-additivity, asymmetrical distribution, dependent of variance on mean. However, when the transformation is done such effects are vanished and these effects are called as scale effects.

Usual procedure of analysis of variance valid under following assumptions

1. **Additive effects:** Treatment effect, block effect and environmental effects are additive.
2. **Independence of errors-** Σ_{ij} N (0, σ^2e)

   This assumption implies that the error of an observation is not related to or dependent upon that of each observation approaches to zero with the variance. This assumption is usually assumed with following proper randomization of the treatments.
3. **Variance is homogenous:** The heterogeneity of variances can be classified due to two reasons.

   I) Variance is functionally related to mean. *e.g.* treatment having low mean has low variance over replications, whereas treatment having high mean has high variance over replications *i.e.* variance mean. In this situation transformation is appropriate remedial measure. By transformation to new scale satisfies the condition of homogeneity of variances.

METHODS OF TRANSFORMATIONS

1. **Logarithmic transformation**:

   This type of transformation is used under following situations.

   (i) When mean and standard deviation are related (If there is significant correlation between mean and SD, then we can conclude that mean and SD are related)

   (ii) When counts are made for wide range of values e.g. No. of insects/plant, No. of eggs/plant

   $Y_{ij} = \log_e (Y_{ij})$; When data is < 10, then transformation should be such that $Y_{ij} = \log (x + 1)$, where, x = No. of observations.

2. **Square root transformation**

   Situation under which it is used

   (i) When mean and variance are related

   (ii) When data consists of small whole number or rare events. *e.g.* Number of infected plants/plot

   (iii) When percent data lies between 0-30 or 70-100 then square root transformation should be carried out.

Mean= $Yij = \sqrt{Yij}$

3. Angular/Sine inverse/Arcsine transformation

Situation under which it is used

When data is on proportions and data obtained of percentages based on count. e.g. Percent non bearing tillers (Which is derived from the ration of the number of non bearing tillers to the total number of tillers).

$Q = \sin^{-1} \sqrt{P}$ *i.e.* $Q\sin^{-2} = P$

When P = 0.5 no transformation is required

Rules for choosing proper transformation scale for percent data derived from count data are

1. For percent data lying within the range of 30-70 percent, no transformation is required.
2. For percent data lying within the range of 0-30 percent or 70-100 percent but not both, square root transformation is required.
3. For percent data lying within the range of 0-100 percent, arcsine transformation is required.

II) Error variance is heterogeneous but there is no relation between variances and real mean

This situation occurs in the experiment conducted in plant breeding trials when treatments are parents, F_1, F_2. Here due to genetic segregation, F_2 population can have high variances than pure line parents. In this case weights which are inversely proportional to variance are considered at the time of analysis.

30 Generation Mean Analysis

The objective of any Crop Breeding is to improve the yield potential of crop by developing new varieties with desirable genetic makeup. However, the yield is a complex character determined by several traits direct selection for which is not effective. Hence, yield can be improved through indirect selection on the basis of yield components. The choice of selection and breeding procedures for genetic improvement of any crop is largely dependent on the knowledge of type and relative amount of genetic component and the presence of non-allelic inter-action for different characters in the plant materials.

Generation mean analysis is one of the quantitative biometric methods based on measurements of phenotypic performances of certain quantitative traits on as many as possible plant individuals in basic experimental breeding generations (parental, filial, backcross and segregation generations). Concept of generation mean analysis was developed by Hayman (1958) and Jinks and Jones (1958) for the estimation of genetic components of variation. Analysis of this technique is based on six different generations of a cross *viz.*, Parents (P_1,P_2), F_1, F_2, and backcrosses (B_1, B_2). The mean values over replications are used for estimation of gene effect.

It is a useful for estimating main gene effects (additive and dominance) of and their digenic (additive × additive, additive × dominance and dominance × dominance) interactions or trigenic (additive × additive × additive, additive × additive × dominance, additive × dominance × dominance and dominance × dominance × dominance) responsible for inheritance of quantitative traits. It also helps in understanding the performance of the parents used in crosses and potential of crosses to be used either for heterosis exploitation or pedigree selection .

Important steps

1. *Season 1:* Selection of parents for crossing: In the first crop season, two parents (Inbred lines, cultivars) having contrasting characters are selected and a cross is made to produce F_1.
2. *Season 2:* F_1 plants of the cross are raised. Some F_1 plants are selfed to obtain F_2.Some of the F1 plants are backcrossed with P_1 and some with P_2 to obtain B_1 and B_2 progeny, respectively.
3. *Season 3:* Evaluation of all six populations in replicated field trial and recording observations on various quantitative characters.

4. Biometrical analysis is done according to the model of generation mean analysis adopted. The biometrical analysis consists of two main steps *viz.*, testing of epistasis and estimation of gene effects and variances.

TESTING OF MODELS (SCALING TESTS)

The testing of presence or absence of epistasis is necessary before the estimation of the different components of genetic variation, because it helps in deciding the method of analysis for the component of variation. The test which determines the presence or absence of epistasis (non allelic interactions) and their type is known as scaling test.

There are two types of scaling tests *viz.*, Individual scaling test and Joint scaling test

I. INDIVIDUAL SCALING TEST

Mather (1949) has given four types of scales *viz.*, A, B, C and D. A number of relations between different generations in absence of epistatic interactions are as under.

$$\bar{B}_1 = \frac{1}{2}(\bar{F}_1 + \bar{P}_1)$$

$$\bar{B}_2 = \frac{1}{2}(\bar{F}_1 + \bar{P}_2)$$

$$\bar{F}_2 = \frac{1}{4}(2\bar{F}_1 + \bar{P}_1 + \bar{P}_2)$$

$$\bar{F}_3 = \frac{1}{4}(2\bar{F}_2 + \bar{P}_1 + \bar{P}_2)$$

This expected relationship is to test the adequacy of the model as below.

$$A = 2\bar{B}_1 - \bar{P}_1 - \bar{F}_1 = 0 \qquad VA = 4V(\bar{B}_1) + V(\bar{P}_1) + V(\bar{F}_1)$$

$$B = 2\bar{B}_1 - \bar{P}_2 - 0 \qquad VB = 4V(\bar{B}_2) + V(\bar{P}_2) + V(\bar{F}_1)$$

$$C = 4\bar{F}_2 - 2\bar{F}_1 - \bar{P}_1 - \bar{P}_2 = 0 \qquad VC = 16V(\bar{F}_2) + 4V(\bar{F}_1) + V(\bar{P}_1) + V(\bar{F}_1)$$

$$D = 2\bar{F}_2 - \bar{B}_1 - \bar{B}_2 = 0 \qquad VD = 4V(\bar{F}_2) + V(\bar{B}_1) + V(\bar{B}_2)$$

When back crosses are absent and F_3 population is available the D scale will be as follow

$$D = 4\bar{F}_3 - 2\bar{F}_2 - \bar{P}_1 - \bar{P}_2 \qquad VD = 16V(\bar{F}_3) + 4V(\bar{F}_3) + V(\bar{P}_1) + V(\bar{P}_2)$$

Then the standard error of A, B, C and D is worked out by taking square root of respective variances and accordingly t values are calculated by the dividing the estimates of A, B, C and D by their respective standard error. If the calculated t values of the scale is greater than 1.96 (*t* table value at 5% level of significance), is considered as significant. The significance of the any one of these scales indicates the presence of non- allelic interactions (Epistasis). It is to be noted that (*i*) The significance of A and B scales indicates the presence of all the three types of non- allelic gene interactions *viz.*, additive x additive, additive x dominance and dominance x dominance.

(*ii*) The significance of C scale suggests dominance x dominance, whereas, the significance of D scale indicates presence of additive x additive type of non-allelic gene interactions and significance of both C and D indicates presence of dominance x dominance and additive x additive types of non- allelic gene interactions.

II. JOINT SCALING TEST

Joint scaling test was proposed by Cavelli (1952) for presence or absence of non- allelic gene interactions. The main drawback of individual scaling tests is that out of six populations only three or four populations are included in the test at a time. However, in joint scaling test any combination of six populations at a time is included. It also provides estimate of three genetic parameters: m, (d), (h) from the mean of all three families available. At least four types of families are required to test the goodness of fit of the model (Weighted least analysis is required). Comparison of these means as observed with expected values derived from the estimates of the three parameters is done by X^2 test. The significance of the chi-square test of different generations indicates the presence of epistasis. The detailed procedure of estimation of joint scaling test is given in solved exercise.

Estimation of Gene Effects

The model to be used for estimation of gene effects are based on the presence or absence of epistatis and on the type of generations included in the study.

I. Three Parameter Model

This model was proposed by Jinks and Jones (1958). This is used in the absence of epistasis as indicated by scaling tests. The gene effects of mean (m), additive (d) and dominance (h) are estimated by the following formulae

$$m = \frac{1}{2}\overline{P}_1 + \frac{1}{2}\overline{P}_2 + 4\overline{F}_3 - 2\overline{B}_1 - 2\overline{B}_2$$

$$(d) = \frac{1}{2}\overline{P}_1 - \frac{1}{2}\overline{P}_2$$

$$(h) = 6\overline{B}_1 + 6\overline{B}_2 - 8\overline{F}_2 - \overline{F}_1 - \frac{3}{2}\overline{P}_1 - \frac{3}{2}\overline{P}_2$$

II. Six Parameter Model

This is used to estimate all the six gene effects viz., mean (*m*), additive (*d*) and dominance (*h*), additive x additive (i), additive x dominance (j) and dominance x dominance (l) in the presence of epistasis as indicated by scaling tests. The gene effects of are estimated by the following formulae proposed by Hayman (1958) and Jinks and Jones (1958).

The formulae proposed by Hayman(1958) are

$m = \overline{F}_2$

$(d) = \overline{B}_1 - \overline{B}_2$

$(h) = \overline{F}_1 - 4\overline{F}_2 - \frac{1}{2}\overline{P}_1 + \frac{1}{2}\overline{P}_2 + 2\overline{B}_1 + 2\overline{B}_2$

$(i) = 2\overline{B}_1 + 2\overline{B}_2 - 4\overline{F}_2$

$(j) = 2\overline{B}_1 - \overline{P}_1 - 2\overline{B}_2 + \overline{P}_2$

$(l) = \overline{P}_1 + \overline{P}_2 + 2\overline{F}_1 + 4\overline{F}_2 - 4\overline{B}_1 - 4\overline{B}_2$

The formulae proposed by Jinks and Jones (1958).

$m = \frac{1}{2}\overline{P}_1 + \frac{1}{2}\overline{P}_2 + 4\overline{F}_2 - 2\overline{B}_1 - 2\overline{B}_2$

$(d) = \frac{1}{2}\overline{P}_1 - \frac{1}{2}\overline{P}_2$

$(h) = 6\overline{B}_1 + 6\overline{B}_2 - 8\overline{F}_1 - \overline{F}_1 - \frac{3}{2}\overline{P}_1 - \frac{3}{2}\overline{P}_2$

$(i) = 2\overline{B}_1 + 2\overline{B}_2 - 4\overline{F}_2$

$(j) = 2\overline{B}_1 - \overline{P}_1 - 2\overline{B}_2 + \overline{P}_2$

$(l) = \overline{P}_1 + \overline{P}_2 + 2\overline{F}_1 + 4\overline{F}_2 - 4\overline{B}_1 - 4\overline{B}_2$

III. Five Parameter Model

This model is used when five generation *viz.*, P_1, P_2, F_1 F_2 and F_3 are available and back cross generations (B_1 and B_2) are not available. In this model, five gene effects viz., mean (*m*), additive (*d*) and dominance (*h*), additive x additive (i), and dominance x dominance (l) in the presence of epistasis as indicated by individual scaling tests (C or D) and joint scaling test. It does not provide information about additive x dominance (*j*) type of non allelic interaction. The five gene effects are estimated by the procedure given by Hayman (1958) as follows

$m = \overline{F}_2$

$(d) = \frac{1}{2}\overline{P}_1 - \frac{1}{2}\overline{P}_2$

$(h) = \frac{1}{6}(4\overline{F}_1 + 12\overline{F}_2 - 16\overline{F}_3)$

$(i) = \overline{P}_1 - \overline{F}_2 + \frac{1}{2}(\overline{P}_1 + \overline{P}_2 + (h) - \frac{1}{2}(l)$

$(l) = \frac{1}{3}(16\overline{F}_3 - 24\overline{F}_2 + 8\overline{F}_1)$

The significance of all the gene effect in all the above model is tested by 't' test. For this, first the standard error of m, (d), (h), (i) (j) and (l) is worked out by taking the squared root of respective variances *viz.*, V_m, $V_{(d)}$, $V_{(h)}$, $V_{(i)}$, $V_{(j)}$, $V_{(l)}$ and 't' values are estimated by dividing the estimated value of m, (d), (h), (i), (j), and (l) by their respective standard error. If the calculated 't' values of the gene effect is greater than 1.96 ('t' table value at 5% level of significance), is considered as significant.

I. Interpretation of results

Significance of a particular gene effect indicates that the inheritance of the character is governed by that gene effect. If more than one gene effects are significant for a particular character, their respective magnitude and direction can be considered. (Greater magnitude of gene effect plays predominant role in the inheritance of that character).

When main effect (*h*) and interaction effect (*l*) present, the + and –ve signs of (*h*) and (*l*) has significance in the expression of the characters. It signs are similar (+ or -), then the non allelic gene interaction is of complementary type which is beneficial for breeding point view. In this case selection is delayed.

It the signs of (*h*) and (*l*) are opposite [+, - or -, +] , the non allelic interaction is duplicate type and under such situation bi-parental mating is to be followed.

If both additive and non additive gene actions are playing role for the inheritance of the character, then go for population improvement approach.

Solved Exercise Individual Scaling tests

A trial was conducted with six generations of a cross in wheat namely P_1, P_2, F_1, F_2, B_1, B_2 replicated twice. The grain yield data recorded on individual plants in each replication of all the six generations.

Step I: Analyzed data for ANOVA and from ANOVA revealed that there were significant differences among the generations and hence proceeded further analysis.

II. Estimates of mean, variances, variances of mean, S.E for six generations

Generation	Mean	Variance	No. of plants	Variance of mean	SE
P_1	20.20	4.38	10	0.44	0.66
P_2	17.40	6.15	10	0.62	0.78
F_1	23.20	7.24	10	0.72	0.85
F_2	20.65	60.50	40	1.51	1.23
B_1	22.20	20.50	20	1.03	1.01
B_2	18.20	21.80	20	1.09	1.04

III. Testing the model by Individual scaling test (Mather, 1949)

Estimates of individual scaling tests as per Mather (1949)

$$A = 2\overline{B}_1 - \overline{P}_1 - \overline{F}_1 = (2 \times 22.20) - 20.20 - 23.20 = 1.00$$

$$B = 2\overline{B}_2 - \overline{P}_2 - \overline{F}_1 = (2 \times 18.20) - 17.40 - 23.20 = -4.20$$

$$C = 4\overline{F}_2 - 2\overline{F}_1 - \overline{P}_1 - \overline{P}_2 = (4 \times 20.65) - (2 \times 23.20) - 20.20 - 17.40 = -1.40$$

$D = 2\bar{F}_2 - \bar{B}_1 - \bar{B}_2 = (2 \times 20.65) - 20.20 - 18.20 = 0.90$

Estimation of variances , SE and 't' value for each scale

Variance of scales		SE		t cal
$V_A = 4V(\bar{B}_1) + V(\bar{P}_1) + V(\bar{F}_1)$	= 5.26	$(V_A)^{1/2}$	= 2.29	= 1.00/2.29 = 0.44
$V_B = 4V(\bar{B}_2) + V(\bar{P}_2) + V(\bar{F}_1)$	= 5.70	$(V_B)^{1/2}$	= 2.39	= – 4.20/2.39 = –1.76
$V_C = 16V(\bar{F}_2) + 4V(\bar{F}_1) + V(\bar{P}_1) + V(\bar{F}_1)$	= 28.15	$(V_C)^{1/2}$	= 5.31	= – 1.40/5.31 = –0.26
$V_D = 4V(\bar{F}_1) + V(\bar{B}_1) + V(\bar{B}_2)$	= 8.17	$(V_D)^{1/2}$	= 2.86	= 0.90/2.86 = 0.31

Scaling test	Estimate	Variance	SE	t cal.
A	1.00	5.26	2.29	0.44
B	-4.20	5.70	2.39	-1.76
C	-1.40	28.15	5.31	-0.26
D	0.90	8.17	2.86	0.31

In this case, all the three scales are found non-significant which indicates absence of non allelic gene interaction hence, the data holds good with additive-dominance model.

Testing the model by Joint scaling test (Cavalli ,1952)

Procedure

1. Three genetic Parameters i.e. *m, d,* and *h* are estimated from the observed means of the three or more generations.
2. Weighted least square analysis is employed (because three parameters are estimated from more than three generation means). Weights are defined as the reciprocal of the squared standard errors and denoted as $1/v(\bar{x})$
3. Three normal equations are required to calculate three parameters (*m, d,* and *h*)

Generation	Mean	Variance	Weight	Coefficient of parameters		
				m	*d*	*h*
P_1	20.20	0.44	2.283	1.0	1.0	0.0
P_2	17.40	0.62	1.626	1.0	-1.0	0.0
F_1	23.20	0.72	1.381	1.0	0.0	1.0
F_2	20.65	1.51	0.661	1.0	0.0	1/2
B_1	22.20	1.03	0.976	1.0	1/2	1/2
B_2	18.20	1.09	0.917	1.0	-1/2	1/2

Equation 1 with respect to (m)

Equation obtained by multiplying each element (weights) by the product of weight and coefficient of m and adding over six generation

Generation	Coff m	wt	cxw	Mean	*m*	*d*	*h*	Mean
P_1	1	2.283	2.283	20.20	2.283	2.283	0.000	46.119
P_2	1	1.626	1.626	17.40	1.626	-1.626	0.000	28.293
F_1	1	1.381	1.381	23.20	1.381	0.000	1.381	32.044
F_2	1	0.661	0.661	20.65	0.661	0.000	0.331	13.653
B_1	1	0.976	0.976	22.20	0.976	0.488	0.488	21.659
B_2	1	0.917	0.917	18.20	0.917	-0.459	0.459	16.697
				Total	**7.845**	**0.686**	**2.658**	**158.464**

Equation 2 with respect to (*d*)

Equation 2 is obtained by multiplying each element (weights) by the product of weight and coefficient of *d* and adding over six generation

Generation	Coff d	wt	cxw	Mean	*m*	*d*	*h*	Mean
P_1	1.0	2.283	2.283	20.20	2.283	2.283	0.000	46.119
P_2	-1.0	1.626	-1.626	17.40	-1.626	1.626	0.000	-28.293
F_1	0.0	1.381	0.000	23.20	0.000	0.000	0.000	0.000
F_2	0.0	0.661	0.000	20.65	0.000	0.000	0.000	0.000
B_1	0.5	0.976	0.488	22.20	0.488	0.244	0.244	10.829
B_2	-0.5	0.917	-0.459	18.20	-0.459	0.229	-0.229	-8.349
				Total	**0.686**	**4.382**	**0.015**	**20.307**

Equation 3 with respect to (h)

Equation 3 is obtained by multiplying each element (weights) by the product of weight and coefficient of *h* and adding over six generation

Generation	Coff h	wt	cxw	Mean	*m*	*d*	*h*	Mean
P_1	0.0	2.283	0.000	20.20	0.000	0.000	0.000	0.000
P_2	0.0	1.626	0.000	17.40	0.000	0.000	0.000	0.000
F_1	1.0	1.381	1.381	23.20	1.381	0.000	1.381	32.044
F_2	0.5	0.661	0.331	20.65	0.331	0.000	0.165	6.826
B_1	0.5	0.976	0.488	22.20	0.488	0.244	0.244	10.829
B_2	0.5	0.917	0.459	18.20	0.459	-0.229	0.229	8.349
				Total	2.658	0.015	2.020	58.049

Now the three equations are as below:

$7.845\ m + 0.686\ d + 2.658\ h = 158.464$ (1)

$0.686\ m + 4.382\ d + 0.015\ h = 20.307$ (2)

$2.658\ m + 0.015\ d + 2.020\ h = 58.049$ (3)

From above three equations, we can estimate the parameters as below

$$\begin{pmatrix} 7.845 & 0.686 & 2.658 \\ 0.686 & 4.382 & 0.015 \\ 2.658 & 0.015 & 2.020 \end{pmatrix} \begin{pmatrix} m \\ d \\ h \end{pmatrix} = \begin{pmatrix} 158.464 \\ 20.307 \\ 58.049 \end{pmatrix}$$

A B C

Say, A × B = C, Hence B= A^{-1} × C

First obtained A^{-1} by adjoin method

Accordingly,

$$\begin{pmatrix} 0.236 & -0.036 & -0.310 \\ -0.036 & 0.234 & 0.046 \\ -0.310 & 0.056 & 0.903 \end{pmatrix} \times \begin{pmatrix} 158.464 \\ 20.307 \\ 58.049 \end{pmatrix} = \begin{pmatrix} 18.622 \\ 1.704 \\ 4.218 \end{pmatrix}$$

A^{-1} C B

SE of each of these estimates is obtained as under root of the diagonal element of inverse matrix (A^{-1})

	Estimate	Variance	SE	t cal
M	18.63**	0.236	0.485	38.364
D	1.70**	0.234	0.483	3.525
H	4.22**	0.903	0.950	4.441

The calculated values of t for all the parameters are highly significant

4. Testing the goodness of fit of the model: Following steps are involved

(*i*) Compute the expected means of these six generations using estimated of (m), (d) and (h) as below.

$\overline{P}_1 = m + d$ = 18.63+1.70 = 20.33

$\overline{P}_2 = m - d$ = 18.63- 1.70 = 16.93

$\overline{F}_1 = m + h$ = 18.63+4.22 = 22.85

$\overline{F}_2$ = m + (1/2h) = 18.63+(1/2 × 4.22) = 20.74

$\overline{B}_1$ = m + (1/2d) + (1/2h) = 18.63 + (1/2 *1.70) + (1/2 × 4.22) = 21.58

$\overline{B}_1^-$ = m - (1/2d) + (1/2h) = 18.63- (1/2 *1.70) + (1/2 × 4.22) = 19.88

Generation	Observed.	Expected	O-E	(O-E)*(O-E)	Weight	X^2
P_1	20.20	20.33	-0.13	0.016	2.283	0.036
P_2	17.40	16.93	0.47	0.232	1.626	0.378
F_1	23.20	22.85	0.37	0.129	1.381	0.178
F_2	20.65	20.74	-0.09	0.007	0.661	0.004
B_1	22.20	21.58	0.62	0.380	0.976	0.371
B_2	18.20	19.88	-1.68	2.820	0.917	2.587
				Calculated	X^2(6-3)df	3.555

The calculated X^2 value 3.555 at 3 df is less than table X^2 value (7.82). Hence, additive dominance model is adequate and there is absence of epistasis interactions.

From both individual scaling and joint scaling test, it is observed that only additive dominant model is adequate for no. of tillers and therefore proceed for three parameter model for estimating m, d, h parameters.

Solved Exercise Three parameter model

As non significance of the A, B, C, and D scales and joint scaling test , only additive dominance model is fitted for the data furnished in above example. Hence, estimates of the genetic parameters by three parameter model (Jinks and Jones, 1958) is carried out

Generation	Mean	Variance	No .of plants	Variance of mean	SE
P_1	20.20	4.38	10	0.44	0.66
P_2	17.40	6.15	10	0.62	0.78
F_1	23.20	7.24	10	0.72	0.85
F_2	20.65	60.50	40	1.51	1.23
B_1	22.20	20.50	20	1.03	1.01
B_2	18.20	21.80	20	1.09	1.04

$$m = \frac{1}{2}\overline{P}_1 + \frac{1}{2}\overline{P}_2 + 4\,\overline{F}_2 - 2\overline{B}_1 - 2\overline{B}_2$$

$$= (0.5 \times 20.20) + (0.5 \times 17.40) + (4 \times 20.65) - (2 \times 22.20) - (2 \times 18.20)$$

$$= 20.60$$

$$(d) = \frac{1}{2}\overline{P}_1 - \frac{1}{2}\overline{P}_2 = (0.5 \times 20.20) - (0.5 \times 17.40) = 1.40$$

$$(h) = 6\overline{B}_1 + 6\overline{B}_2 - 8\overline{F}_2 - \overline{F}_1 - \frac{3}{2}\overline{P}_1 - \frac{3}{2}\overline{P}_2$$

$$= (6 \times 22.20) + (6 \times 18.20) - (8 \times 20.65) - (23.20) - (1.5 \times 20.20 - (1.50 \times 17.40)$$

$$= -2.40$$

Test of significance

$V_m = 1/4V(\overline{P}_1) + 1/4\ (\overline{P}_2) + 16V(\overline{F}_2) + 4V(\overline{B}_1) + 4V(\overline{B}_2)$

$= (1/4 \times 0.44) + (1/4 \times 0.62) + (16 \times 1.51) + (4 \times 1.03) + (4 \times 1.09) = 32.92$

$SEm = \sqrt{32.92} = 5.73$

$t = \frac{20.60}{5.73} = 3.59$

Table t at 99 df = 1.98

Hence, *m* is significant

$V_d = 1/4V(\overline{P}_1) + 1/4V(\overline{P}_2) = (1/4 \times 0.44) + (1/4 \times 0.62) = 0.26$

$SEd = \sqrt{0.26} = 0.51$

$t = \frac{1.40}{0.51} = 2.73$

Table t at 19 df = 2.093

Hence, *d* is significant

$V_h = 36V(\overline{B}_1) + 36V(\overline{B}_2) + 64V(\overline{F}_2) + V(\overline{F}_1) + 9/4(\overline{P}_1) + 9/4(\overline{P}_2)$

$= (36 \times 1.03) + (36 \times 1.09) + (64 \times 1.51) + (0.72) + (9/4 \times 0.44) + (9/4 \times 0.62)$

$= 176.03$

$SEh = \sqrt{176.03} = 32.27$

$t = \frac{-2.40}{32.27} = -0.18$

Table *t* at 109 df = 1.98; the calculated *t* is < than table t at 5 % level, hence, *h* is not significant

CONCLUSION

From the three parameter model, it is revealed that residual effect (m) and additive component (d) are found significant and positive, hence the character is governed by additive gene action and hence to exploit the additive gene effects simple pedigree selection method is suggested for improvement number of tillers.

Solved Exercise: Six parameter model

Estimates of mean, variances and variances of mean of six generations of a cross of pearl millet inbreeds for ear head girth are given below. Estimate the different components as per six parameter model (Hayman, 1958).

Generation	Mean	Variance	No. of plants	Variance of mean	SE
P_1	11.05	0.48	10	0.048	0.22
P_2	9.06	0.35	10	0.035	0.19
F_1	10.85	0.33	10	0.033	0.18
F_2	10.55	2.84	40	0.071	0.27
B_1	11.75	1.96	20	0.098	0.31
B_2	10.69	0.99	20	0.049	0.22

Estimates of individual scaling tests as per Mather (1949)

Scaling test	Estimate	Variance	SE	t cal.
A	1.60*	0.47	0.69	2.33
B	1.47*	0.27	0.51	2.86
C	0.39	1.35	1.16	0.34
D	-1.34*	0.43	0.66	-2.04

Joint scaling test: X^2 (6-3) df = 12.63 **

Both individual scaling test and joint scaling test are significant which indicate the presence of non allelic interaction and hence components of genetic parameters are estimated by following six parameter model by Hayman (1958).

Estimation of Six parameters as suggested by [Hayman, 1958)

$$m = \overline{F}_2 = 10.55$$

$$(d) = \overline{B}_1 - \overline{B}_2 = (11.75 - 10.69) = 1.06$$

$$(h) = \overline{F}_1 - 4\overline{F}_2 + \frac{1}{2}\overline{P}_1 - \frac{1}{2}\overline{P}_2 + 2\overline{B}_1 + 2\overline{B}_2$$

$$= (10.85) - (4 \times 10.55) - (0.5 \times 11.05) - (0.5 \times 9.06) + (2 \times 11.75) + (2 \times 10.69)$$

$$= 3.47$$

$$(i) = 2\overline{B}_1 + 2\overline{B}_2 - 4\overline{F}_2 = (2 \times 11.75) + (2 \times 10.69) - (4 \times 10.55) = 2.68$$

$$(j) = 2\overline{B}_1 - \overline{P}_1 - 2\overline{B}_2 + \overline{P}_2 = (2 \times 11.75) - (11.05) - (2 \times 10.69) + (9.06) = 0.06$$

$$(l) = \overline{P}_1 + \overline{P}_2 + 2\overline{F}_1 + 4\overline{F}_2 - 4\overline{B}_1 - 4\overline{B}_2$$

$$= (11.05) + (9.06) + (2 \times 10.85) + (4 \times 10.55) - (4 \times 11.75) - (4 \times 10.69) = -5.75$$

For testing the significance estimates of variances, SE and t of six parameters are computed as below.

$$Vm = V(\overline{F}_2) = 0.071$$

$$V(d) = V(\overline{B}_1) + V(\overline{B}_2) = (0.098 + 0.049) = 0.147$$

$$V(h) = V(\overline{F}_1) + 16V(\overline{F}_2) + \frac{1}{4}V(\overline{P}_1) + \frac{1}{4}V(\overline{P}_2) + 4V(\overline{B}_2) + 4V(\overline{B}_2)$$

$= (0.033) + (16 \times 0.071) + (0.25 \times 0.048) + (0.25 \times 0.035) + (4 \times 0.98) +$

$(4 \times 0.049) = 1.777$

$V(i) = 4V(\bar{B}_1) + 4V(\bar{B}_2) + 16V(\bar{F}_2)$

$= (4 \times 0.098) + (4 \times 0.049) + (16 \times 0.071) = 1.723$

$V(j) = 4V(\bar{B}_1) + V(\bar{P}_1) + 4(\bar{B}_2) + V(\bar{P}_2)$

$= (4 \times 0.098) + (0.048) + (4 \times 0.049) + (0.035) = 0.168$

$V(l) = V(\bar{P}_1) + V(\bar{P}_2) + 4V(\bar{F}_1) + 16V(\bar{F}_2) + 16V(\bar{B}_1) + 16V(\bar{B}_2)$

$= (0.048) + (0.035) + (4 \times 0.033) + (16 \times 0.071) + (16 \times 0.098) + (16 \times 0.049)$

$= 3.704$

Estimation of SE and t of parameters.

SE (*m*) =	$(Vm)^{1/2}$	=	0.266		t cal (m)	=Est. (*m*)/SE (*m*)	39.621
SE(*d*) =	$(Vd)^{1/2}$	=	0.384		t cal (d)	=Est. (d)/SE (*d*)	2.762
SE (*h*) =	$(Vh)^{1/2}$	=	1.333		t cal (h)	=Est. (*h*)/SE (*h*)	2.607
SE (*i*) =	$(Vi)^{1/2}$	=	1.313		t cal (i)	=Est. (*i*)/SE (*i*)	2.041
SE (*j*) =	$(Vj)^{1/2}$	=	0.410		t cal (j)	=Est. (*j*)/SE (*j*)	0.159
SE (*l*) =	$(Vl)^{1/2}$	=	1.925		t cal (l)	=Est. (*l*)/SE(*l*)	-2.988

Parameter	**Est.**	**SE**	**t cal**
m	10.55*	0.266	39.621
d	1.06**	0.384	2.762
h	3.475*	1.333	2.607
i	2.68*	1.313	2.041
j	0.065	0.410	0.159
l	-5.75**	1.925	-2.988

CONCLUSION

The genetic parameter (*m*), (*d*), (*h*) and interaction effect (i) and (l) are found to be significant. However, the (i) type gene effect has higher magnitude and in desirable direction and hence additive x additive gene effect plays predominant role in the inheritance of that character. However, the opposite sign of (*h*) and (*l*) gene effects indicates presence of duplicate epistasis, hence it is suggested biparental mating in segregation generation for further improvement of earhead girth in pearl millet

Solved Exercise : Five parameter analysis

Estimates of mean, variances, variances of mean of five generations (P_1, P_2, F_1, F_2 and F_3) of a cross in cotton for test weight /plant are given below. Test the epitasis interaction by following individual scaling test and estimate the different components of genetic variation as per five parameter model (Hayman, 1958).

Generation	Mean	Variance	No. of plants	Variance of mean
P_1	12.30	0.120	15	0.008
P_2	11.40	0.800	15	0.053
F_1	12.00	1.300	15	0.087
F_2	11.20	12.450	60	0.208
F_3	10.55	6.500	40	0.163

Estimates of individual scaling tests as per Mather (1949)

Formulae	Estimate	Variance	SE	t cal.
$C = 4\bar{F}_2 - 2\bar{F}_1 - \bar{P}_1 - \bar{P}_2$	19.90**	3.73	1.93	10.31
$D = 4\bar{F}_3 - 2\bar{F}_2 - \bar{P}_1 - \bar{P}_2$	20.40**	3.57	1.89	10.80

Estimation of five parameters as suggested by Hayman (1958)

$m =$	$\bar{F}_2$	=	11.20
$d =$	$\frac{1}{2}\bar{P}_1 - \frac{1}{2}\bar{P}_2$	=	0.45
$h =$	$\frac{1}{6}(4\bar{F}_1 + 12\bar{F}_2 - 16\bar{F}_3)$	=	2.27
$i =$	$\bar{P}_1 - \bar{F}_2 + \frac{1}{2}(\bar{P}_1 + \bar{P}_2 + (h)) - \frac{1}{2}(l)$	=	2.85
$l =$	$\frac{1}{3}(16\bar{F}_3 - 24\bar{F}_2 + 8\bar{F}_1)$	=	–1.33

Estimation of variances of parameters

$Vm = V\bar{F}_2$	= 0.208
$Vd = \frac{1}{4}V(\bar{P}_1) + \frac{1}{4}V(\bar{P}_2)$	= 0.015
$Vh = \frac{1}{36}(16V(\bar{F}_1) + 144V(\bar{F}_2) + 256V(\bar{F}_3)$	= 2.040
$Vi = V(\bar{P}_1) + V(\bar{F}_2) + \frac{1}{4}V(\bar{P}_1 + \bar{P}_2 + (h)) + \frac{1}{4}V(l)$	= 1.897
$Vl = \frac{1}{9}(256V(\bar{F}_3) + 576V(\bar{F}_2) + 64V(\bar{F}_1))$	= 18.500

Estimates of Five parameters [Hayman, 1958)

Parameter	Est.	SE	t cal
m	11.20**	0.456	24.587
d	0.45**	0.124	3.634
h	2.27	1.428	1.586
i	2.85*	1.377	2.069
l	–1.33	4.301	–0.310

Conclusion

The genetic parameter main effect (*d*) and interaction effect (*i*) are found to be significant. However, the additive *x* additive (*i*) type gene effect has higher magnitude and in desirable direction. Hence, it is suggested to follow the pedigree selection in segregation generation for further improvement of test weight in cotton.

THREE PARAMETER MODEL (FOR NON-INTERACTING CROSSES)

In the absence of the epitasis, A Mather (1949) estimated three components of variances viz., Additive (D), Dominance (H) and environmental (E) variances. The analysis is based on the variances of different generations.

$$E = \frac{V_{P1} + V_{P2} + V_{F1}}{3}$$

$$D = 2(V_{F2} - V_{B1} - V_{B2})$$

$$H = 4(V_{F2} - \frac{1}{2}VD - E)$$

Degree of dominance can also be estimated by formulae $\left(\frac{H}{D}\right)^{1/2}$

If the ratio is equal to one, there is complete dominance, if it is greater than one, it indicates over dominance. If the ratio is greater than zero but less than one, it indicates partial dominance and if the ratio is zero, there is no dominance.

SOLVED EXERCISE

The variances of six generations viz., P1, P2, F1, F2 , B1 and B2 for grain yield of a cross is given below. Partition the variance components as proposed by Mather (1949)

Generation	Variance	Generation	Variance
P_1	4.38	F_2	60.50
P_2	6.15	B_1	20.50
F_1	7.24	B_2	21.80

$$\text{Environmental variance (E)} = \frac{V_{P1} + V_{P2}V_{F1}}{3} = \frac{(4.38 + 6.15 + 7.24)}{3} = 5.92$$

$D = 2(V_{F2} - V_{B1} - V_{B2}) = 2 \times (60.50 - 20.50 - 21.80) = 36.40$

$H = 4(V_{F2} - \frac{1}{2}VD - E) = 4 \times [(60.50 - (1/2 - 36.40) - 5.92] = 145.52$

Degree of dominance can also be estimated

Degree of dominance = $(H/D)^{1/2} = (145.52/36.40)^{1/2} = 2.00$

CONCLUSION

The dominant component was very high than additive component and the dominance ratio is also more than one. Hence, the character is governed by dominant gene action. To exploit the dominance gene action, heterosis breeding is suggested.

31 Heterosis and Inbreeding Depression

HETEROSIS

The term of heterosis was coined by Shull (1914). Heterosis or hybrid vigour refers to the superiority in performance of hybrid individuals in one or more characters over its parents. In other words, heterosis refers to increase of hybrid in fitness and vigour over the parental values. In current usage, heterosis and hybrid vigour are used as synonyms and interchangeable. However, by the original definition, heterosis refers to the developmental stimulation resulting from the union of different gaments (cause), whereas hybrid vigour denotes the manifest effect of heterosis *i.e.* the phenotypic expression (effect) of the genetic phenomenon. Hybrid cultivars are used for commercial production of number of plant species. They are a desirable type of cultivar because of their ability to capitalize on heterosis.

Measurement of Heterosis

The performance of a hybrid relative to its parent can be expressed in two ways. While, heterosis can be classified into three types on the basis of estimation.

(i) **Mid parent heterosis or relative /average heterosis :** It is the performance of a hybrids compared with the average performance of its parents.

(ii) **Better parent heterosis or Heterobeltosis :** It is a comparison of the performance of the hybrid with that of best parent in the cross.

(iii) **Standard heterosis:** The superiority of F_1 hybrid over standard commercial variety/hybrid is known as standard heterosis. The is also called as useful or economic heterosis. This type of heterosis has practical value in plant breeding.

Heterosis usually is expressed as percent and computed as follow

$$\text{Mid parent heterosis (\%)} = \frac{\overline{F}_1 - \overline{MP}}{\overline{MP}} \times 100$$

$$\text{Better parent heterosis (\%)} = \frac{\overline{F}_1 - \overline{BP}}{\overline{BP}} \times 100$$

$$\text{Economic heterosis (\%)} = \frac{\overline{F}_1 - \overline{SV}}{\overline{SV}} \times 100$$

$$\text{Standard heterosis (\%)} = \frac{\overline{F}_1 - \overline{SH}}{\overline{SH}} \times 100$$

Where,

$\overline{F_1}$ = Mean value of F1 hybrid

$\overline{MP}$ = Mean value of average of two parents involved in the cross

$\overline{BP}$ = Mean value of better parent involved in the cross

$\overline{SV}$ = Mean value of standard variety used as check

$\overline{SH}$ = Mean value of standard hybid used as check

Testing of significance of heterosis: by t test

Standard error

Relative heterosis = $\sqrt{\frac{3}{2r}EMS}$

Heterobeltosis and standard heterosis = $\sqrt{\frac{2}{r}EMS}$

Where,

EMS = Error mean square

r = Number of replications

t value for relative heterosis = $\frac{\overline{F_1}-\overline{MP}}{SE}$

t value for heterobeltosis = $\frac{\overline{F_1}-\overline{BP}}{SE}$

t value for standard heterosis = $\frac{\overline{F_1}-\overline{SV}}{SE}$

Compare the calculated t value with table t value at error degree of freedom.

Or

The significance of inbreeding depression is tested with the help of critical difference as below

Critical difference CD = Standard error difference × t value at 5 % or 1%.

The difference between F1 and parent used for estimation of heterosis (Mid parent, better parent or standard commercial variety) is taken into account cross wise.When the difference between mean value of F_1 and parent is greater than CD or 't' calculated is greater than 't' table value, it is considered as significant.

Genetic Basis of Heterosis

There are three possible genetic cause of heterosis *viz.*, dominance, over dominace and epistasis

(*i*) **Dominance hypothesis:** This theory was proposed by Davenport (1908), Bruce (1901) and Keeble and Pellew (1910). According to this hypothesis, heterosis is the result of superiority of dominant alleles when recessive alleles are deleterious. Here deleterious recessive genes of one parent are hidden by the dominant genes of the another parent and hybrid exhibit heterosis. This is the most widely accepted explanation of heterosis.

There are two objections to dominance hypothesis. First, if the hypothesis is true, it should be possible to obtain pure heterotic individuals in F_2 which are homozygous for all dominant gene. Jones (1971) provided explanation for this and proposed dominance of linked gene to explain heterosis. He suggested that there are many linkage between some favourable dominant genes and some unfavourable recessive genes and as result it is not possible to obtain true breeding homozygous individual for all dominant genes in F_2 generation.

The second objection is that if the heterosis is due to dominance, the F_2 curve should be skewed towards dominant genes, but the curve of F_2 is found always smooth and symmetrical. Collins (1921) provided explanation for this objection that characters like yield is governed by polygenes which exhibit continuous variation resulting symmetrical distribution of genes.

(ii) Overdominance hypothesis: This theory was independently proposed by Shull and East in 1908 and supported by East (1936) and Hull (1945). According to this hypothesis, heterosis is the result of superiority of heterozygote over its both homozygous parents. Thus, heterosis is directly proportional to the heterozygosis.The superiority of heterozygote may arise either due to production of superior hybrid substance in heterozygote which is completely different from either of the homozygous product or due to greater buffering capacity in the heterozygote resulting from the cumulative action of divergent alleles or stimulation of divergent alleles. Hence this theory is called various names such as stimulation of heterozygosis, cumulative action of divergent alleles, single gene heterosis, super dominance and overdominance.

(iii) Epistasis: It ie well established that the occurrence and magnitude of heterosis has positive association with presence and magnitude of non allelic interaction i.e epistasis. Particularly, dominance × dominance non allelic interaction may contribute to heterosis.

Genetic Factors Affecting Heterosis

There are four major genetic factors which affect magnitude of heterosis in crop plants.

(i) Mode pollination: The level of heterosis is generally higher in cross pollinated species than n self pollinated species.

(ii) Genetic diversity of parents: Maximum heterosis occurs at an optimal or intermediate level of parental diversity.

(iii) Genetic base of parents: The higher heterosis occurs when parents have broad genetic base.

(iv) Adaptability of parents: In many crops, heterosis is associated with wide adaptability of the parents.

Inbreeding Depression

The mating between closely related individual is known as inbreeding. The closest form of inbreeding is self fertilization; other forms are sib matings and half sib matings.

Inbreeding depression refers to decrease in fitness and vigour due to inbreeding. The degree of inbreeding is measured by the inbreeding co–efficient. The main features of inbreeding depression are (i) Inbreeding results from mating between closely related individuals and (ii) Inbreeding depression results due to fixation of of unfavourable recessive genes in $F_{2.}$

Estimation of Inbreeding Depression

The inbreeding depression is estimated when both F_1 and F_2 populations of the same cross are available.

$$\text{Inbreeding Depression (\%)} = \frac{\overline{F}_1 - \overline{F}_2}{\overline{F}_1} \times 100$$

Where,

$\overline{F}_1$ = Mean value of F_1 hybrid

$\overline{F}_2$ = Mean value of F_2

The significance of inbreeding depression is tested with the help of t' test or critical difference as below

$$\text{Standard error} = \sqrt{\frac{2}{r}\text{EMS}}$$

$$\text{t value for Inbreeding depression} = \frac{\overline{F}_1 - \overline{F}_2}{\text{SE}}$$

Or

The significance of inbreeding depression is tested with the help of critical difference as below

Critical difference CD = Standard error difference × t value at 5 % or 1%.

The difference between F_1 and F_2 mean is taken into account cross wise. When the difference between mean value of F_1 and F_2 is greater than CD or 't' calculated is greater than 't' table value, it is considered as significant.

Inferences

The following inferences about the type of gene action involved in the expression of the quantitative charaters can be drawn from the estimates of the heterosis and inbreeding depression.

(*i*) If high heterosis is followed by high inbreeding depression, it indicates presence of non additive gene action.

(*ii*) If high heterosis is followed by low inbreeding depression, it indicates presence of additive gene action.

(*iii*) If high heterosis is is negative in F_1 and there is increase in F_2, it also indicates presence of additive gene action.

(*iv*) The genes with lack of dominance will not exhibit heterosis in F_1 but may show increase in performance in F_2 (Low inbreeding depression) due to fixation of genes *i.e.* additive gene action.

SOLVED EXERCISE

A total of 9 hybrids (F_1s), 9 hybrids (F_2s), and their 6 parents and one standard check hybrid were evaluated for grain yield in a RBD with three replications. The analysis of variance for RBD is carried out and following are the mean values over replications for F1s, F2s, parents and standard commerciall hybrid. From the data worked out heterosis percent (all types) and inbreeding depression as follows.

The following is meandata on grain yield (g). Estimated the heterosis percent (three types) as follows

S. N.	F_1s	Mean F_1s	Mean F_2s		Parents	Mean Parents
1	P_1xP_4	67.00	67.50		P_1	61.17
2	P_2xP_4	93.27	91.27		P_2	90.70
3	P_3xP_4	111.67	120.00		P_3	97.10
4	P_1xP_5	97.00	85.00		P_4	58.47
5	P_2xP_5	108.00	90.00		P_5	94.37
6	P_3xP_5	109.13	115.00		P_6	84.77
7	P_1xP_6	91.00	81.00		Check hybrid	97.10
8	P_2xP_6	92.67	92.70			1445.33
9	P_3xP_6	91.93	93.00			

ANOVA table

Source	df	SS	MS	.F
Replication	2	99.59	49.80	3.61
Treatments	24	17552.18	731.34	53.07
Error	48	661.53	13.78	
Total	74	18313.30		

Estimation of heterosis percent:

For $P_1 \times P_4$

$$\text{Relative Heterosis (\%)} = \frac{\overline{F_1} - \overline{MP}}{\overline{MP}} \times 100 = \frac{67.00 - 59.82}{59.82} \times 100 = 12.00$$

$$SE = \sqrt{\frac{3}{2r} EMS} = \sqrt{\frac{3}{6} \times 13.78} = 2.62$$

$$\text{t value for relative heterosis} = \frac{\overline{F_1} - \overline{MP}}{SE} = \frac{7.18}{2.62} = 2.74$$

Better parent heterosis (%) = $\frac{\overline{F}_1 - \overline{BP}}{\overline{BP}} \times 100 = \frac{67.00 - 61.67}{61.67} \times 100 = 9.53$

$SE = \sqrt{\frac{2}{r} EMS} = \sqrt{\frac{2}{3} \times 13.78} = 3.03$

t value for relative heterosis = $\frac{\overline{F}_1 - \overline{BP}}{SE} = \frac{5.83}{3.03} = 1.92$

Standard heterosis (%) = $\frac{\overline{F}_1 - \overline{SH}}{\overline{SH}} \times 100 = \frac{67.00 - 97.10}{97.10} \times 100 = -31.00$

$SE = \sqrt{\frac{2}{r} EMS} = \sqrt{\frac{2}{3} \times 13.78} = 3.03$

t value for relative heterosis = $\frac{\overline{F}_1 - \overline{SH}}{SE} = \frac{-30.10}{3.03} = -9.93$

Similarly, for all hybrids heterosis pecent and t values were worked out and presented in table.

t table values are 1.99 and 2.64 at 74 df at at 5 % and 1% level of significance.

Inbreeding Depression

For $\mathbf{F_2 (P_1 \times P_4)}$

Inbreeding Depression (%) = $\frac{\overline{F}_1 - \overline{F}_2}{\overline{F}_1} \times 100 = \frac{67.00 - 67.501}{67.00} \times 100 = -0.75$

The significance of inbreeding depression is tested with the help of t' test

$SE = \sqrt{\frac{2}{r} EMS} = \sqrt{\frac{2}{3} \times 13.78} = 3.03$

t value for Inbreeding depression = $\frac{\overline{F}_1 - \overline{F}_2}{SE} = \frac{-0.50}{3.03} = -0.17$

Similarly, inbreeding depression for other crosses was worked out and presented in table.

Percent Heterosis and inbreeding depression for grain yield in nine crosses

Hybrid	Relative heterosis		Heterobeltiosis		Standard heterosis		Inbreeding depression	
	(%)	t cal.	(%)	t cal.	(%)	t cal.	(%)	t cal.
$P_1 \times P_4$	12.00**	2.74	9.53	1.92	–31.00**	–9.93	–0.75	–0.17
$P_2 \times P_4$	25.05**	7.13	2.83	0.85	–3.94	–1.26	2.14	0.66
$P_3 \times P_4$	43.56**	12.93	15.01**	4.81	15.01**	4.81	–7.46**	–2.75
$P_1 \times P_5$	24.73**	7.34	2.79	0.87	–0.10	–0.03	12.37**	3.96
$P_2 \times P_5$	16.71**	5.90	14.44**	4.50	11.23**	3.60	16.67**	5.94
$P_3 \times P_5$	13.99**	5.11	12.39**	3.97	12.39**	3.97	–5.38	–1.94
$P_1 \times P_6$	24.71**	6.88	7.35*	2.06	–6.28*	–2.01	10.99**	3.30
$P_2 \times P_6$	5.62	1.88	2.17	0.65	–4.56	–1.46	–0.03	–0.01
$P_3 \times P_6$	1.09	0.38	–5.32	–1.71	–5.32	–1.71	–1.16	–0.35

CONCLUSION

Heterosis

The hybrids $P_3 \times P_4$, $P_2 x P_5$ and $P_3 \times P_5$ recorded significantly positive relative heterosis, heterobeltiosis and standard heterosis.

Inbreeding depression

The highest significant inbreeding depression was observed in case of the crosses $P_2 \times P_5$ followed by $P_1 \times P_5$ and $P_1 \times P_6$. In case of the hybrid $P_2 \times P_5$ high heterosis was followed by high inbreeding depression indicating presence of non –additive gene action, whereas, the hybrid $P_3 \times P_4$ had heterosis and low inbreeding depression, indicated presence of additive gene action.

32 Significance of Gene Action

The gene actions governing a polygenic trait are grouped as additive (fixable variation) and non– additive (Unfixable variation). The non –additive gene action further classified into dominance and epistatic interactions. The epistatic interaction is further sub–divided into additive × additive, additive × dominance and dominance × dominance.

The implication of these types of gene action indifferent breeding programmer are as below

Mode of pollination	**Type of gene action**	**Breeding procedure to be followed**
Self pollinated crops	Additive	Pure line selection, mass selection, progeny selection and hybridization followed by selection with pedigree breeding
	Non –additive	Heterosis breeding and recombination breeding with postponement of selection at later generations..
Cross pollinated crops	Additive	Synthetic breeding, composite breeding and population improvement by recurrent selection for gca.
	Non –additive	Heterosis breeding an population improvement by recurrent selection for sca.
	Both additive and non additive	Population improvement by Reciprocal recurrent selection.

Further the implications of all the type of gene action in two major breeding programmes *viz.*, recombination breeding and heterosis.

Breeding programme	**Type of gene action**	
Recombination Breeding	Additive	Chance of fixing by selection is easy to improve the trait
	Non –additive	Likely to confuse selection, since it may not breed true. It will reduce the value by 50 in the next generation. Postpone the selection to later generation (6–8) by reduction of heterozygosity
	Additive × Additive Interaction (i)	Fixe in homozygous condition easily by selection

Contd...

Contd...

	Additive × Dominance Interaction (j)	Partial fixable
	Dominance × Dominance Interaction (l)	Non–fixable. In cross pollinated crops, this can be exploited by intercrossing and selecting the individuals based on their performance (Recurrent selection)

Heterosis Breeding

The additive, dominance and additive × additive and dominance × dominance interactions are more important in consideration with the expression of heterosis. In case of additive dominance model expected magnitude of heterosis is calculated by

$$F_1 - P_1 = (m + [d]) - (m - [d]) \text{ Since } F_1 = m + [h] \text{ and } P_1 = m + [d]$$
$$= m + [h] - m - [d]$$
$$= [h] - [d]$$

For positive heterosis to occur $[h]$ must be positive and greater in magnitude than $[d]$.

This $[h] > [d]$ in turn requires one or both of two conditions

(*i*) $h > d$ for some or all the genes *i.e.* over dominance at some or all loci.

(*ii*) $d < \Sigma d$: *i.e.* there must be dispersion of genes between the parents. If heterosis is due to dispersion of genes, recombination breeding is best to produce a true breeding line expressing the character to atleast as high a degree as the F_1.

When digenic interaction is present, the magnitude of heterosis is calculated as

$$F_1 - P_1 = (m + [h]) + [l] - m - [d] - (i)$$
$$\text{As, } F_1 = m + [h] + [l] \text{ and } P_1 = m + [d] + [i]$$
$$= [h] + [l] - [d] - [i]$$

For positive heterosis to occur $[h] + [l]$ must be positive and greater in magnitude than $[d] + [i]$. In presence of digeneic non–allelic interaction heterosis more likely to arise, and that to of greater magnitude if the signs of [h] and [l] are the same directions (complementary type) and the magnitude of $[d]$ and $[i]$ is smaller.

33 Stability Analysis

GENOTYPE × ENVIRONMENT INTERACTION

The phenotype of an individual is determined by the effects of its genotype and the environment surrounding it. The effects of genotype and environment on phenotype may not be always independent. The phenotypic response to change in environment is not the same for all genotypes, the consequences of variation in phenotype depend upon the environment.

Successful cultivars must have good yield and other essential agronomic characters. Besides, their performance should be reliable over a wide range of environmental conditions. The basic cause of differences in stability between genotypes is a wide occurrence of genotype × environment interactions (G × E). It is therefore the interplay between genetic and non–genetic effects on development of genotype (Comstock and Moll, 1963). G × E interaction causes fluctuations of yield across environments. In other words, G × E is a differential genotypic expression across environments.

G × E interaction is of major consequence to breeders in the process of developing improved varieties. When varieties are grown at several locations for testing their performance, their relative rankings usually do not remain the same. This causes difficulty in demonstrating significant superiority of any variety. G × E interaction is present whether varieties are pure lines, single crosses, double crosses, top–crosses, S1 lines or any other material with which the breeder is working (Dabholkar, 1999).

An understanding of environmental and genotypic causes of G × E interaction is important at all stages of plant breeding, including ideotype design, parent selection based on

traits, and selection based on yield (Jackson *et al.*, 1998). Understanding of the causes of G × E interaction can be used to establish breeding objectives, to identify ideal test conditions, and to formulate recommendations for areas of optimal cultivar adaptation. It can also help to reduce the cost of extensive genotype evaluation by eliminating unnecessary testing sites and by fine tuning the breeding programmes. The presence of a large G × E interaction may necessitate establishment of additional testing sites, thus increasing the cost of developing commercially important varieties (Kang, 1996).

Genes and Environment

Organisms are determined neither by their genes nor by their environment; they are the consequence of the interaction of genes and environment (Suzuki *et al.*, 1981). Genotype describes the complete set of genes inherited by an individual that is important for the expression of a trait under investigation. Phenotype describes all aspects of the individual's morphology, physiology and ecological relationships. The genotype is essentially a fixed character of the organism; it remains constant throughout life and is unchanged by environmental effects. The phenotype changes continually and the direction of that change is a function of the sequence of environments that the individual experiences (Suzuki *et al.*, 1981).

The sum total of the effects of physical, chemical and biological factors of an individual other than its genotype is known as the environment. Comstock and Moll (1963) classified environments into two categories,

(*i*) **Macro–environment** : The environment which is associated with a given location or area at a particular period of time.

(*ii*) **Micro-environment:** The environment of a single organism as opposed to that of another organism growing at the same time and in almost the same place. It includes physical and chemical attributes of soil, climatic variables, solar radiation, insect pests and disease. The macro-environments reflect a collection of micro–environments which are more alike within each macro–environment with the result that macro–environments substantially differ from each other.

The terms 'predictable and unpredictable environments' were coined by Allard and Bradshaw (1964) to define and classify environments.

(*i*) **Predictable environment:** The predictable environment includes the regular and more or less permanent features of the environment such as climate as determined by its longitude and latitude, soil type, rainfall and day length. It also includes what are called controllable variables (Perkins and Jinks, 1971) e.g. the level of fertilizer applied, sowing date and sowing density, amount of irrigation and others that can be artificially created.

(*ii*) **Unpredictable environment:** The unpredictable or uncontrollable environments, on the other hand, include weather fluctuations such as differences between seasons in terms of amount and distribution of rainfall and the prevailing temperature during the crop growth. The absence or low level of interaction will be useful for uncontrollable variables, whereas for the controllable variables a high level of interaction in the favourable direction is desirable to obtain maximal performance (Chahal and Gosal, 2002).

Significance of Genotype × Environment Interaction

The breeding programmes aimed to develop stable genotypes also depend upon whether a breeder is dealing with predictable or unpredictable environmental variation. Whenever dealing with predictable environmental variation, the first step that should be taken is to identify the differences. There is no difficulty when differences are recognizable, for example, differences in the seasons such as varieties to be developed for the rainy season or post–rainy season. Breeders can develop varieties suitable for both these seasons because environmental variation is defined.

For variety trials, which are tested in the same locations (L) and genotypes (G) and over years (Y), G × E analysis of variance may be partitioned into components due to G × L, G × Y and G × L × Y. Significance of mean square for G × L generally suggests that the region for which genotypes are being bred comprises of a number of special environments. In such circumstances the geographic region could be subdivided into sub regions which are relatively homogeneous. Varieties should be bred which are specifically adapted to these ecotypes. Implication of G × Y interaction is very different from G × L interaction. This is so because year to year fluctuations cannot be predicted in advance and breeders can hardly aim their programmes to develop varieties suited to particular years (Dabholkar, 1999).

In some situations, environmental variation is predictable but can also be corrected. For example, saline soils can be corrected by certain agronomic practices or by addition of some amendments. This is easier and quicker than evolving varieties suitable for such situations. However, breeding of varieties suitable for saline or acidic soils is low cost input and also a relatively permanent solution to the problem. It is relatively easier to develop varieties specifically adapted to predictable environmental situations than to breed for unpredictable environmental variations. The aim of the breeding programme should, therefore be to develop genotypes that can withstand unpredictable transient environmental fluctuations. In other words, breed widely adapted genotypes (Dabholkar, 1999).

According to Allard and Bradshaw (1964) "a variety which can adjust its genotypic or phenotypic state in response to transient fluctuations in environment in such a way that it gives high and stable economic returns for place and year, is termed as well buffered". Plant breeders generally agree that the new variety must show a high degree of stability in performance.

The concept of stability

Two basic concepts of phenotypic stability are distinguished as (i) the biological concept, and (ii) the dynamic concept.

(*i*) **The biological concept of stability**: It refers to the constant performance of a genotype over a wide range of environments. According to Becker and Léon (1988) in static stability a genotype posses unchanged performance regardless of variation of the environments, thus implying that its variance among environments is zero. This type is seldom a desired feature of crop cultivars, since no response to improved growing conditions would be expected.

(*ii*) **Dynamic stability or agronomical concept of stability:** It implies that a stable genotype should always give high yield expected at the level of productivity of the respective environments, *i.e.*, a variety with G × E interaction as small as possible (Becker, 1981; Dabholkar, 1999). With quantitative traits, the majority of genotypes often react similarly to favourable or unfavourable environmental conditions.

Adaption and adapabilty:

Adaptation is the property of a genotype which permits its survival under selection. According to Simmonds (1962) adaptation has four separable aspects. These are:

(*i*) **Specific genotypic adaptation:** It is close to adaptation of the corresponding genotypes to a limited environment.

(*ii*) **General genotypic adaptation:** It is the capacity of a genotype to produce a range of phenotypes adapted to a variety of environments.

(*iii*) **Specific population adaptation:** It is analogous to (1) and is the aspect of specific adaptation of heterogeneous population that is attributable to interaction between components rather than to the adaptations of components themselves.

(*iv*) **General population adaptation:** It is analogous to general genotypic adaptation and is the capacity of a heterogeneous population to adapt to a variety of environments.

A genotype maintains high yield over a wide range of environments, it is referred to as having general or wider adaptation. On the other hand, if this is true only for a limited range of environments, that genotype has specific or narrow adaptation. Wider adaptability or stability is the result of homeostasis, which refers to the buffering capacity to genotype to environmental fluctuations (Lerner,1954). The homeostasis is of two types; (1) Genetic homeostasis– Genetic buffering capacity of a genotype to environmental fluctuations and physiological homeostasis– Physiological or development buffering capacity.

Steps in Stability Analysis

Stability analysis is done from the data of replicated rials conducted over several environments (Locations/years etc). The stability analysis consists of following steps.

1. ***Environment–wise analysis of Variance:*** Following usual method of analysis of variance, the data are analysied for a quantitative traits in all environments separately. The data of environments, where significant difference for geneotypes are observed, are used for pooled analysis.
2. ***Pooled analysis of variance:*** A two way table is formed for tabulating the data of different environments as under.

Genotype	E_1	E_2	——	E_n
1				
2				
.				
.				
n				

$E_{1,}$ E_2__ E_n are the eman data over replications in different environments. From this two way table, the pooled analysis of variance is constructed.

Pooled Analysis of Variance

Source	**Df**	**SS**	**MS**
Genotype	g–1		$\sigma_e^2 + r\sigma_{ge}^2 + re\sigma_g^2$
Environment	e–1		$\sigma_e^2 + r\sigma_{ge}^2 + re\sigma_e^2$
G × E interaction	(g–1) (e–1)		$\sigma_e^2 + r\sigma_{ge}^2$
Pooled error	e(r–1) (g–1)		σ_e^2
Total	r(ge–1)		

$$\text{MS due to pooled error} = \frac{\text{SS in } E_1 + \text{SS in } E_2 + \text{...SS in } E_n}{\text{Error df in rror df in } E_1 + \text{...error df in } E_n}$$

The MS due to pooled error is required to test the significance of G × E interaction. If the G × E interaction is found significant, the analysis may be proceeded further for estimating the phenotypic stability. A number of models are known for for estimation of phenotypic stability.

Models for Stability Analysis Analysis

The statistics, which can be used to identify stable genotypes, are classified into parametric and non–parametric.

I. **Parametric (empirical and statistical one):** It is more common and involves relating observed genotypic responses, in terms of yield, to a sample of environmental conditions. It is useful when the data are continuous. The parametric stability statistics is further divided into two Groups as follows:

A. Conventional models

1. Stability factor model:

Lewis (1954) suggested simple measure of phenotypic stability which he termed "stability factor' and expressed as:

$$S.F. = \frac{\overline{X}_{H.E.}}{\overline{X}_{L.E.}}$$

where, $\overline{X}$ is the mean value and H.E. and L. E. are high and low yielding environments, respectively.

The greater the S.F > deviates from unity, the less stable is the genotype.

2. Ecovalence Model–

Wricke (1962) proposed using the contribution of each genotype to the G × E interaction sum of squares as a stability measure and defined this concept or statistics as ecovalence (*Wi*).

Ecovalence is simple to calculate and is expressed as:

$$Wi = Si\,(Yij - \overline{Yi} - \overline{Yj} + \overline{Y}..)^2$$

where, *Yij* is the mean performance of genotype *i* in the *jth* environment and Yi. and Y.j are the genotype and environment mean deviations respectively, and *Y* .. is the overall mean. For this reason, genotypes with a low *Wi* value have smaller deviations from the overall mean across environments and are thus more stable. According to the meaning of the ecovalence, this stable genotype possesses a high ecovalence (low values of *Wi* = high ecovalence).

3. Mean variance component of pairwise G × E interaction:

Plaisted and Peterson (1959) described a procedure to characterize the stability of the performance of several varities. A combined analysis of variance at all locations was computed for each pair of varieties g (g –1)/2 pairs for g varieities and an estimate of $\sigma^2 gl$ was obtained for each pair and for each variety. The variety with the smallest mean value was considered the most stable and so on.

4. **Stability variance Model :**

Shukla (1972) defined the stability variance of genotype *i* as its variance across environments after the main effects of environmental means have been removed. Since the genotype main effect is constant, the stability variance is thus based on the residual (GE*ij*+ e ij) matrix in a two way classification. The stability statistic is termed "stability variance" (σ_i^2) and is estimated as follows:

$$\sigma_i^2 = \frac{1}{(G-1)(G-2)(E-1)}\left[G(G-1)\Sigma_i\left(Yij - \overline{Y}_j + \overline{Y}..\right)^2 - \Sigma_i\Sigma_j\left(Yij - \overline{Yi.} - \overline{Y.j} + \overline{Y.}\right).^2\right]$$

where *Yij* is the mean yield of the ith genotype in the jth environment, *j Y*. is the mean of all genotypes in j^{th} environments and *Y* .. is the mean of all genotypes in all environments. A genotype is called stable if its stability variance (σ_i^2) is equal to the environmental variance. As the stability variance is the difference between two sums of squares, it can be negative, but negative estimates of variance are not uncommon in variance component problems. Negative estimates of (σ_i^2) may be taken as equal to zero as usual (Shukla, 1972).

5. Coefficient of determination (ri^2)

Pinthus (1973) proposed to use the coefficient of determination (ri^2) instead of deviation mean squares to estimate stability of genotypes, because ri^2 is strongly related to S^2di.

$r_i^2 = 1 - \frac{S_{di}^2}{S_{xi}^2}$ both ri^2 and *bi* has the advantage of being dependent of units of measurement.

6. Lin and Binns model :

Lin and Binns (1988) defined the superiority measure (Pi) of the ith genotype as the mean square of distance between the ith genotype and the genotype with the maximum response as:

$$Pi = \frac{\left[n(Yi. - M..)^2 + \left(Yij - Yi + M_j + Mii\right)^2\right]}{2n}$$

where Y_{ij} is the average response of the ith genotype in the jth environment, *Yi.* is the mean deviation of genotype *i, Mj* is the genotype with maximum response among all the genotypes in the *jth* locations, and *n* is the number of locations. The smaller the value of *Pi,* the less is the distance to the genotype with maximum yield and the better the genotype.

B. Regression coefficient models

Regression coefficient (*bi*) and deviation mean square (S^2di)

The regression of each genotype's mean yield against the mean yields of an environment is determined and the stability range is determined by the main effects multiplied by the regression coefficients of genotypes. The G × E interaction is divided into two segments i) a component due to linear regression (*bi*) of the ith genotype on the environment mean and ii) a deviation (dij) : GEij= *bi*Ej + dij

Therefore, Yij= μ + Gi + Ej + (*bi*Ej + dij) + eij

1. Finlay and Wilkinson (1963) :

Finlay and and Wilkinson (1963) used two parameters , namely (i) mean performance over environments and (ii) regression of performance in different environments over the respective environmental mean. In this model the deviation from the regression line are not estimated which are important for stability analysis.

2. Eberhart and Russell model (1966)

Eberhart and Russel (1966) made improvement in stability analysis by portioning the genotypic environment interaction of each variety into to parts *viz.,* (i) slope of the regression line , and (ii) deviation from th regression line. In this model, total variance is first divided into two components *i.e.* genotypes and environment + interaction (E + Gx E). The second component (E + G × E) is further sub divided into two components *viz.,* (a) environment linear, (b), G × E linear and (C) pooled deviations. The sum of squares due to pooled deviations are further divided into SS due to individual genotype. This model consists of three parents *viz.,* mean over locations/seasons., regression coefficient (bi) and deviation from regression (S^2di).

3. Perkins and Jinks model (1968)

In Perkins and Jinks (1968) model, the total variance is first divided into three components, *viz.*, genotypes, environments and genotypes × environments. The G × E variance is subdivided into (a) heterogeneity due to regression, and (b) sum of square due to reminder. The SS reminder further divided into *SS* due to individual genotype. This model also includes three parameters *viz.*, mean over locations/ seasons., regression coefficient (bi) and deviation from regression (S^2di).

4. Freeman and Perkins model (1971)

In Freeman and Perkins (1971) model, the total variance is first divided into three components, *viz.*, genotypes, environments and genotypes × environments. The environmental SS is sub divided into two components, namely (a) combined regression, and (b) residual 1. The interaction variance is also sub divided into two parts, viz. (a) homogeneity of regression, and (b) residual 2. The two residual SS form the pooled deviations in this model. This model also includes three parameters viz., mean over locations/seasons., regression coefficient (bi) and deviation from regression (S^2di).

In regression models, the marginal means of the environments is used as independent variables in the regression analysis and the interaction is restricted to a multiplicative form. Finlay and Wilkinson (1963) defined a genotype with *bi* = 0 as stable (static concept) and Eberhart and Russell (1966) defined a genotype with *bi* = 1 as stable (dynamic concept). Many scientists consider *bi* as a response parameter and S^2 di as a stability parameter.

Regression coefficients approximating 1.0 indicate average stability. When this is associated with high mean yield, varieties have good general adaptability. When associated with low mean yield, genotypes are poorly adapted to all environments. Regression values above 1.0, describe genotypes with increasing sensitivity to environmental change (below average stability) and greater specificity of adaptability to high yielding environments. Regression coefficients below 1.0 provide a measure of greater resistance to environmental change (above average stability) and, therefore, increasing specificity of adaptability to low yielding environments.

5. Genotypic stability model –Tai (1971)

Tai (1971) uses two statistics as in the case of Eberhart and Russel (1966) , *viz.*, Linear response of a genotypes to environmental effects (α_i) and deviation from linear response (λ_i). Stable genotype (α_I, λ_i) (–1, 1).

Limitations of regression approach

(*i*) Genotype means are not independent of site means.

(*ii*) Genotype that has a consistent response to environmental factors, but is a typical of the genotypes in the trial, will have a high deviation from regression.

(*iii*) Does not consider underlying causes for performance at each site . Sites with deficiencies and excesses of water, nutrients, etc. will be juxtaposed on the regression line, resulting in large deviations from regression.

(*iv*) Heritabilities of b_i and s_d^2 are low.

(*v*) Stability is a statistical, not a genetic concept

The stability concept by Becker and Léon (1988), Lin *et al.* (1986) categorized stability in to three types:

Type I : If the among–environment variance of a genotype is small, the genotype is considered to be stable. This concept is useful for quality traits, disease resistance or for stress characters. According to this concept a genotype performs the same in different environments or under different environmental conditions. This stability is static or can be seen as a biological concept of stability. Genotype variances across environments (Si^2) and the coefficient of variability (CV_i) are used as parameters to describe this type of stability (Francis and Kannenburg, 1978).

Type II : A genotype is considered to be stable if its response to environments is parallel to the mean response of all genotypes in the trial. This concept is called the dynamic or agronomic concept of stability. In this case, a stable genotype has no deviations from the general response to environments and creates a possible way of predicting the response of a genotype to a certain environment. Parameters used to describe this type of stability are regression coefficients (*bi*) (Finlay and Wilkinson, 1963) and Shukla's (1972) stability variance (σ_i^2)

Type III : A genotype is considered to be stable if the residual mean square from the regression model on an environmental index is small. The environmental index is the mean yield of all the genotypes in each location minus the grand mean of all the genotypes in all locations. The method of Eberhart and Russell (1966) and Tai (1971) can be used for estimating type III stability.

Although many authors and breeders used the regression approach, simultaneous studies emphasized the limitations.

C. Non–parametric techniques for stability analysis

Non–parametric (analytical clustering) approach defines environments and phenotypes in terms of biotic and abiotic factors and is useful when the data are discontinuous. Non–parametric data analysis has the potential to reduce complex data into intuitive measures of stability. When sample size is small non parametric method is the obvious choice, unless the nature of the population is exactly known. In non paramtric methods genotypes are ranked within environments based on the yield performance. The use of this method generally provides stability of rank of genotypes across the environment. Broadly it is classified in two parts (i). Detection of G × E interaction and (ii) Stability by non parametric indices. The various types of nap paramtric measures are as follows

a. Based on ranks [Huehn (1979), Nasar and Huehn (1987) and Huhen and Nassar (1989)]
b. Kang rank sum method (Kang,1988)

c. Modified rank sum by Kang (1991)

d. Yield stability Static (Kang,1993).

e. Non Parametric indices (Eskridge,1990)

Advantages of non parametric stability static compared to parametric measures (Huehn, 1990a)

(*i*) Reduction of even avoidance of the bias caused by outliers.

(*ii*) Assumption free *i.e.* no assumption are needed about the distribution of data ((Normality, additivity and homogeneity of error variance))

(*iii*) Stability parameters based on ranks are easy to interpret

(*iv*) When large number of genotypes are tested, the power of test of NP test is very close to parametric test.

(*v*) The use of non parametric methods becomes inevitable when the parametric method fails to provide the valid interpretations due to the presence of large non linear G × E interaction.

D. Multivariate analysis techniques

Multivariate techniques are widely applied in stability analysis to provide further information on real multivariate response of genotypes to environments. The multivariate analysis has three main purposes: 1) to eliminate noise from the data pattern, 2) to summarize the data, 3) to reveal the structure in the data (Becker and Léon, 1988). Through multivariate analysis, genotypes with similar responses can be clustered, hypothesized, and later tested, and their data can be easily summarized and analyzed (Crossa, 1990 and Gauch, 1982a).

Multivariate analyses are appropriate for analyzing two–way matrices of G genotypes and E environments. The response of any genotype in E environments may be conceived as a pattern in E–dimensional space, with the coordinate of an individual axis being the yield or other metric of the genotype in one environment. Crossa (1990) has also distinguished two groups of multivariate techniques used to elucidate the internal structure of G × E interactions:

1. **Ordination techniques:** Principal component analysis, principal coordinate analysis and factor analysis, assume that the data are continuous. These techniques attempt to represent G × E relationships as faithfully as possible in a low dimensional space. A graphical output displays similar genotypes or environments close to each other and dissimilar items farther apart. Ordination is effective for showing relationships and reducing noise (Gauch, 1982b).
2. **Classification techniques:** Cluster analysis and discriminant analysis, seek discontinuities in the data. These methods involve grouping similar entities in clusters and effective for summarizing redundancy in the data (Crossa, 1990).

A. Principle component Analysis

i. Perkins (1972), Freeman and Dowker (1975), Seif *et al.*, 1979)

Salient features:

1. PCA is a multivariate approach that provides an alternative to conventional multiple regression analysis. PCA transforms the data into linear combinations of the original variables that are uncorrelated.
2. Useful for determining which agro climatic or biotic factors can be used to discriminate among environments.
3. The principal component analysis was found to be efficient in describing G × E interactions.

Drawback for PCA – interpretation is not as intuitive as with multiple regression or correlation analyses. Results do not bear any obvious relationship to environmental variables.

(*ii*) Principal coordinate analysis

Principal coordinate analysis is a generalization to the PCA in which any measure of similarity between individuals can be used. Its objectives and limitations are similar to those of PCA. Crossa (1990) highlighted some of the advantages:

(*i*) it is trustworthy when used for data that include extremely low or high yielding sites;

(*ii*) it does not depend on the set of genotypes included in the analysis;

(*iii*) and it is simple to identify stable varieties from the sequence of graphic displays.

(*iii*) Additive Main Effects and Multiplicative Interaction Model (AMMI)

(Gauch, 1988; Freeman, 1990 and Zobel ,1990)

Salient features:

1. Method for analyzing GEI to identify patterns of interaction and reduce background noise
2. Combines conventional ANOVA with principal component analysis.
3. May provide more reliable estimates of genotype performance than the mean across sites.
4. Biplots help to visualize relationships among genotypes and environments; show both main and interaction effects.
5. Enables to identify target breeding environments and to choose representative testing sites in those environments.
6. Enables to select varieties with good adaptation to target breeding environments.
7. Can be used to identify key agroclimatic factors, disease and insect pests, and physiological traits that determine adaptation to environments.
8. A type of fixed effect, Linear–Bilinear Model.

(*iv*) Shifted multiplicative model (Cornelius *et al.*,1992)

Salient features:

1. Search for "**separability**" of crop cultivars based on GEI .
3. Identify clusters of genotypes in which crossover interactions are negligible.
4. Cluster environments so that the genotypes within each group show no crossover interactions.

(*v*) Factorial Regression (FR)

Salient features

1. A fixed effect, linear model.
2. Can incorporate additional genotypic and environmental covariables into the model.
3. Similar to stepwise multiple regression, where additional variables are added to the model in sequence until sufficient variability due to GEI can be explained.
4. FR is easier to interpret than PLS, but may give misleading results when there are correlations among the explanatory variables in the model.

(*vi*) Linear–Bilinear Mixed Models

Salient features

1. Have become widely accepted for analysis of GEI.
2. Lead to Factor Analytic form of the genetic variance–covariance for environments.
3. Has desirable statistical properties.
4. When genotypes are random, co ancestries can be accommodated in the model.

(*vii*) GGE or SREG (Sites Regression) Model

Salient features:

1. Fixed effect, linear–bilinear model that is similar to AMMI.
2. Only the environmental effects are removed before PCA.
3. The bilinear term includes both the main effects of genotype and GXE effects.
4. May be used to evaluate test environments (Yan and Holland, 2010).

(*viii*) Partial Least Squares Regression (PLS)

Salient features

1. AMMI model only considers one response variable.
2. PLS is a type of bilinear model that can utilize information about environmental factors (covariables) – rainfall, temperature and soil type.
3. PLS can accommodate additional genotypic data – disease reistance, molecular marker scores.

4. Analysis indicates which environmental factors or genotypic traits can be used to predict GEI for grain yield.

B. Cluster analysis : Westcot (1987) and Webber and Wricke (1990)

Cluster analysis is a numerical classification technique that defines groups or clusters of individuals. Two types of classification can be distinguished. The first is non–hierarchical classification, which assigns each item into a class. The second type is hierarchical classification, which groups individuals into clusters and arranges these into a hierarchy for the purpose of studying relationships in the data.

Salient features

1. Stratify environments into homogeneous groups so that there is no GEI within groups; maximize between group variation
2. Possible benefits:

   May suggest genetic relationships

   May lead to better understanding of environmental factors influencing adaptation

   Permits systematic approach for choosing testing sites
3. Disadvantage – location groupings are seldom consistent from one year to the next
4. Many choices of criteria exist for calculating distance beween sites (Euclidian distance, dissimilarity index) and assigning groups (e.g., average linkage or UPGMA)

C. Pattern Analysis Cooper and DeLacy, (1994)

Salient features: Steps involved

(*i*) Recommended pretreatment (transformation) – scale the data by removing environment main effects and adjust scale by dividing by the phenotypic standard deviation at each site.

(*ii*) Use a classification procedure to identify environments which show similar discrimination among the genotypes.

(*iii*) Use an ordination procedure (singular value decomposition) – similar to AMMI except that it uses transformed data

(*iv*) Use biplots to show relationships between genotypes and environments

(*v*) This method groups genotypes and environments according to either their similarity (ordination method) or dissimilarity (classification methods).

D. Factor analysis (Johnson and Wichern (1982) and Calinski et al., (1987)

Factor analysis is also related to PCA. The variables of the factor analysis are similar to the components of the latter. In this procedure, a large number of variables are reduced to a small number of main factors. Variation is explained in terms of general factors common to all variables and in terms of factors unique to each variable.

I. EBERHART AND RUSSEL MODEL

Eberhart and Russel (1966) in their stability analysis portioned the genotype–environment interaction of each genotype into two parts namely (i) slope of the regression line and (ii) deviation from the regression line. They used the following model to study the stability of genotypes under different environments.

$$Yij = m + Bi\ Ij + \delta ij$$

Where, Yij = is the mean of the ith genotype in j^{th} environment.

M = is the mean of all varieties over all the environments,

Bi = is the regression coefficient of ith variety on environmental index that measures the response of ith genotype to the varying environments,

Ij = Environmental index *i.e.* the deviation of the mean of all varieties at a given environment from the overall mean.

δij = The deviation from regression of the ith variety at j^{th} environment.

Analysis of Variance

The pooled analysis of variance can be further extended in which the total mean squares may be portioned into various components. In this model, the total variance is first divided into two components, *i.e.* (i) genotype and environment plus interaction (E + (G × E. The second component is further sub–divided into three components viz., (a) environment linear, (b) G × E linear and (c) pooled deviations. The variance due to pooled deviations are further divided into variance due to individual genotype.

Analysis of Variance for Stability

Source of Variation	df	Sum of Square	MS
Total	(ge–1)	$\Sigma\Sigma Y_{ij}^2 - CF$	
Genotypes	(g–1)	$\frac{\Sigma Y_{i.}^2}{e} - CF$	MS_1
E + (G × E)	g(e–1)	$\Sigma\Sigma Y_{ij}^2 - \frac{\Sigma Y_{.i}^2}{e}$	
Environment (linear)	1	$\frac{(\Sigma Y_{ij} I_j)^2}{g/\Sigma I_j^2}$	
G × E (linear)	(g–1)	$\Sigma\left[\frac{(\Sigma YijI_j^2)}{\Sigma I_j^2}\right] - \left[\frac{(\Sigma\ YjI_j^2)}{g/\Sigma I_j^2}\right]$	MS_2
Pooled Deviation	g(e–2)	$\Sigma\Sigma S_{ij}^2$	MS_3

Contd...

Contd...

Genotype 1 . . Genotype g	(e–2) . . (e–2)	$\left[\Sigma Y_{ij}^2 - \frac{Y_{.i}^2}{e}\right] - \left[\frac{(\Sigma YjI_j)^2}{\Sigma I_j^2}\right]$	
Pooled Error	e (r–1) (g–1)	Pooled replications × genotype SS over environments	Ms_4=Me′

F test

1. To test the significance of differences among the genotype means $F = \frac{MS_1}{MS_3}$
2. To test whether the varieties differ for their regression on the environmental index

   $F = \frac{MS_2}{MS_3}$
3. To test individual deviation from linear regression

   $$F = \frac{\left[\frac{\delta^2_{ij}}{(e-2)}\right]}{\text{Pooled error}}$$

Stability parameters

(*i*) The regression co–efficient (bi), which is the regression of the performance of each genotype under different environments on the environmental means over all the genotypes.

$$b_i = \frac{\Sigma Y_{ij} I_j}{\Sigma I_j^2}$$

where,

$$I_j = \frac{\Sigma Y_{ij}}{g} - \frac{\Sigma\Sigma Y_{ij}}{ge}$$

(*ii*) Mean square deviation (s^2d) from linear regression

$$S^2d = \frac{\Sigma\delta^2_{ij}}{e-2} - \frac{S_e^2}{r}$$

Where,

$$\Sigma\delta^2_{ij} = \left[\Sigma Y_{ij}^2 - \frac{Y_i^2}{g}\right] - \left[\frac{(\Sigma Y_{ij} I_i)^2}{\Sigma I_j^2}\right]$$

S^2e = Estimate of pooled error

Eberhart and Russel (1966) defined a stable variety as one having high mean yield with a regression coefficient of unity (b=1) and a minimum deviation from the regression line (S^2_d =0). Regression coefficient (bi) is considered as parameter of response and S^2d as the parameters of stability.

Stable genotype

1. A genotype with high/desirable mean value
2. A genotype with deviation not significantly deviating from zero is said to be stable.

3. (*a*) A genotype with unit regresiion coefficient (b=1)–not significantly deviating from

   1 is said to be average responsive *i.e.* suitable for all environments.

   (*b*) A genotype with significant b value (b>1)– is said to be highly responsive *i.e.* suitable for favourable environments.

   (*c*) A genotype with significant b value (b<1)– is said to be low responsive *i.e.* suitable for unfavourable environments.

SOLVED EXERCISE

A total of 16 genotypes of Niger were evaluated in three environments in RBD with three replications. The following is the mean data over environments on seed yield (kg /plot) . Worked out the stability parameters for different genotypes and interpret the results.

Step I: Analysis of variance for individual environment: Since in all the three environments the MS due to genotypes are significant, all the three environments are used for pooled analysis.

Step; II Prepare two way table

Sr. No.	Genotypes	Seed yield (kg/plot)				
		E_1	E_2	E_3	Total	Mean
1	G_1	14.32	27.68	15.78	57.78	19.26
2	G_2	25.80	39.20	22.24	87.24	29.08
3	G_3	22.59	39.06	25.00	86.65	28.88
4	G_4	11.98	28.65	16.96	57.59	19.20
5	G_5	12.41	27.28	22.56	62.25	20.75
6	G_6	12.28	21.62	13.38	47.28	15.76
7	G_7	12.59	32.28	16.91	61.78	20.59
8	G_8	12.96	31.98	12.84	57.78	19.26
9	G_9	15.31	28.26	18.0	61.57	20.52
10	G_{10}	17.10	37.62	17.1.	71.82	23.94
11	G_{11}	8.910	35.11	17.63	61.65	20.55
12	G_{12}	13.95	35.99	16.52	66.46	22.15
13	G_{13}	13.95	30.86	20.52	65.33	21.78
14	G_{14}	10.43	24.23	12.95	47.61	15.87
15	G_{15}	14.69	27.54	14.7.	56.93	18.98
16	G_{16}	31.48	44.68	30.78	106.94	35.65
	Total = $Y_{.j}$	250.75	512.04	293.87	1056.66	352.22
	Mean = Y.1,.2,.3	15.67	32.00	18.37	66.04	22.01

III. Pooled Analysis

$$CF = \frac{(GT)^2}{g \times e} = \frac{(352.22)^2}{48} = 23261.05$$

Table SS = $14.32^2 + \ldots + 30.78^2 - CF = 3914.94$

$$\text{Genotype SS} = \frac{(57.78)^2 + ... + (106.94)^2}{3} - CF = 1211.78$$

$$\text{Environment SS} = \frac{(250.75)^2 + (512.04)^2 (293.87)^2}{16} - CF = 2452.71$$

G × E interaction SS = Table SS–Genotype Ss–Environment SS
= 3914.94 –1211.78 – 2452.71= 250.46

$$\text{MS due to pooled error SS} = \frac{\text{SS in } E_1 + \text{SS in } E_1 .... \text{SS in } E_n}{\text{Error } df \text{ in } E_1 + \text{rror } df \text{ in } E_1 .... \text{rror } df \text{ in } E_n}$$

$$= \frac{487.45}{90} = 5.42$$

Pooled Analysis of Variance

	df	SS	MSS	F
Genotype	15	1211.78	80.79**	44.64
Environment	2	2452.71	1226.36**	677.55
G × E	30	250.46	8.35**	4.61
Pooled error	90	487.45	5.42/3=1.81	
Total	143			

The pooled analysis was done using the mean data of 3 replications . Thus MS due to pooled error = 5.42/3=1.81

Since, the G × E interaction is significant, the analysis may be proceeded for further for estimating stability parameters.

Eberhart and Russel Model

1. Estimation of Environmental Index (I_j)

$$I_j = \frac{\sum_j Yij}{g} - \frac{\Sigma_i \Sigma_j Yij}{ge}$$

$$I_1 \text{ for } E_1 = \frac{250.75}{16} - \frac{1056.66}{48} = -6.34$$

$$I_2 \text{ for } E_2 = \frac{512.04}{16} - \frac{1056.66}{48} = 9.99$$

$$I_3 \text{ for } E_3 = \frac{293.87}{16} - \frac{1056.66}{48} = -3.64$$

2. Regression Coefficient (bi for each variety)

$$bi = \frac{\sum_j Y_i I_j}{\sum_j I_j^2}$$

$\Sigma Y_{ij} I_j$ for genotype 1= (14.32 × – 6.34) + (27.68 × 9.99) + (15.78 × –3.64) =128.13

ΣI_j^2 for all genotype = $(-6.34)^2 + (9.99)^2 + (-3.64)^2$ = 153.25

b_1 for all genotypes is estimated as follows =

Genotype	$\sum_i Y_{ij} I_j / \sum_j I_j^2$	b_i	Genotype		b_i
G_1	128.13/153.25	0.84	G_9	119.54/153.25	0.78
G_2	146.83/153.25	0.96	G_{10}	204.97/153.25	1.34
G_3	155.73/153.25	1.02	G_{11}	229.90/153.25	1.50
G_4	148.35/153.25	0.97	G_{12}	210.78/153.25	1.37
G_5	111.52/153.25	0.73	G_{13}	144.95/153.25	0.95
G_6	89.28/153.25	0.58	G_{14}	128.65/153.25	0.84
G_7	180.92/153.25	1.18	G_{15}	128.32/153.25	0.84
G_8	190.42/153.25	1.24	G_{16}	134.40/153.25	0.88

Note mean of b = 1.00

3. Mean square deviation (s^2d) from linear regression for each variety

$$\overline{S}^2_d = \frac{\Sigma \delta^2_{ij}}{e-2} - \frac{S^2 e}{r}$$

Where,

$$\Sigma \delta^2_{ij} = \left[\Sigma Y_{ij}^2 - \frac{Y_i^2}{e}\right] - \left[\frac{(\Sigma Y_{ij} I_i)^2}{\Sigma I_j^2}\right]$$

S^2e/r = MS due to pooled error

Genotype 1:

$$\Sigma Y_{ij}^2 - \frac{Y_i^2}{e} = (14.32^2 + 27.68^2 + 15.78^2) - (57.78)^2/3 = 1220.25 - 112.84 = 107.41$$

$$\frac{(\Sigma Y_{ij} I_i)^2}{\Sigma I_j^2} = \frac{[(14.32x - 6.34) + (27.68 \times 9.99) + (15.78\ x - 3.64)]^2}{153.25}$$

$$= \frac{16417.30}{153.25} = 107.13$$

$$\Sigma \delta_{ij}^2 = 107.41 - 107.13 = 0.28$$

$$\overline{S}^2_d = \frac{0.28}{1} - 1.81 = -1.53$$

Similarly, $\Sigma \delta_2 ij$ and $\overline{S}^2_d$ values for all the genotypes were estimated and given as below

Geno-type	$\Sigma\delta^2_{ij}$	$\bar{S}^2_d$		Genotype	$\Sigma\delta^2_{ij}$	
G_1	0.28	–1.53		G_9	0.16	–1.65
G_2	19.28	17.47		G_{10}	6.57	4.76
G_3	0.00	–1.81		G_{11}	11.12	9.31
G_4	2.84	1.03		G_{12}	0.58	–1.23
G_5	34.32	32.51		G_{13}	8.24	6.43
G_6	0.10	–1.71		G_{14}	0.01	–1.8
G_7	0.61	–1.20		G_{15}	2.55	0.74
G_8	6.10	4.29		G_{16}	4.78	2.97

Computation of variances due to different sources

S.S. due to environment + (variety × environment) = $\Sigma Y_{ij}^2 - \frac{Y_i^2}{e}$

= 27175.99–24472.83 = 2703.16

S.S. due to environment (linear) =

= SS due to environment from pooled analysis of variance=2452.71

S.S. due to variety × environment (Linear) = $\sum i\left[\frac{\Sigma_j Y_{ij} I_j^2}{\Sigma i I_j^2}\right]$

= 2605.62 – 2452.71 = 152.91

S.S. due to pooled deviation

= S.S. due to environment + (variety × environment)– S.S. due to env. (l)– S.S. due to V × E (l)

= 2703.16 – 2452.71 – 152.91 = 97.54

S.S due to pooled deviation for individual genotype is computed and given in ANOVA for stability table.

ANOVA for Stability as per Eberhart and Russel model

Source	df	SS	MSS	Cal F
Genotype	15	1211.78	80.79* (MS_1)	13.24
Environment + (Geno.x Env.)	32	2703.16	84.47	13.84
Env. Linear	1	2452.71	2452.71*	402.08
Geno. × Env. (linear)	15	152.91	10.19* (MS_2)	1.67
Pooled deviation	16	97.54	6.10* (MS_3)	3.37
G_1	1	0.28	0.28	0.16
G_2	1	19.28	19.28*	10.65
G_3	1	0.00	0.00	0.00
G_4	1	2.84	2.84	1.57
G_5	1	34.32	34.32*	18.96

Contd...

Contd...

G_6	1	0.10	0.10	0.06
G_7	1	0.61	0.61	0.34
G_8	1	6.10	6.10	3.37
G_9	1	0.16	0.16	0.09
G_{10}	1	6.57	6.57*	3.63
G_{11}	1	11.12	11.12*	6.14
G_{12}	1	0.58	0.58	0.32
G_{13}	1	8.24	8.24*	4.55
G_{14}	1	0.01	0.01	0.01
G_{15}	1	2.55	2.55	1.41
G_{16}	1	4.78	4.78*	2.64
Pooled error	90		1.81 (MS_4)	

Analysis of variance of stability showed highly significant differences among genotypes, environments and genotypes × environments interaction. Highly significant difference in the genotypes under study may be due to differences in their genetic make up and diverse nature of origin. The highly significant differences in locations indicated an existence of variation in the prevailing environments. . Highly significant differences are very important for determining G × E interaction. Relatively higher value of the linear component as compared to nonlinear one, suggested the possibility of prediction of performance for yield over the environments.

Stability parameters

(*i*) **High/desirable mean :** The genotype having mean greater than grand mean

$$SE_m = \sqrt{\frac{\text{MS due to pooled deviation}}{e-1}} = \sqrt{\frac{6.10}{2}} = 1.75$$

$$CD = \mathbf{SE_m} \times \sqrt{2} \times \text{t at pooled error df} = 1.75 \times 1.414 \times 1.99 = 4.92$$

The CD may be used to decide about the superiority of mean of individual genotypes.

(*ii*) **Deviation from linear regression ($\overline{S}^2_d = 0$)**

The $\overline{S}^2_d$ all genotypes are already presented in table.

To test the deviation of individual genotypes from regression , the F test is used and estimates and significance are already presented in stability ANOVA.

(*iii*) **Regression coefficient**

The regression coefficient (bi) for all the genotypes are already presented. To test the significance of bi values the following test is used .

$$SE_{bi} = \sqrt{\frac{\text{MS due to pooled deviation of the ith genotype}}{\Sigma I_j^2}}$$

Where, MS due to pooled deviation of ith genotype = $\frac{\delta^2_{ij}}{e-2}$

$$\mathbf{SE_{b1}} = \sqrt{\frac{0.28}{153.25}} = 0.04$$

$$t = \frac{b_1 - 1}{SEb1} = \frac{0.84 - 1}{0.04} = 3.5$$

Compare t calculated value with t table at g (e–2) df.

Similarly the significance is worked out for all the genotypes.

Estimates of Stability parameters as per Eberhart and Russel Model

Gen.	**Mean**	**bi**	**S²di**
G1	19.26	0.84**	–5.10
G2	29.08	0.96	13.90**
G3	28.88	1.02	–5.36
G4	19.20	0.97	–2.54
G5	20.75	0.73	28.93**
G6	15.76	0.58**	–5.31
G7	20.59	1.18*	–4.76
G8	19.26	1.24	0.74
G9	20.52	0.78**	–5.24
G10	23.94	1.34	1.23
G11	20.55	1.50	5.79**
G12	22.15	1.38**	–4.76
G13	21.78	0.95	2.86
G14	15.87	0.84**	–5.39
G15	18.98	0.84	–2.84
G16	35.65	0.88	–0.62
G.M	**22.01**		
SE (bi)		**0.80**	

CONCLUSION

The genotypes G–1, G–3, G–4, G–6, G–7, G–8, G–9, G–10, G–12, G–13, G–14, G–15 and G–16 deviated non significantly from zero (S^2di =0). Hence, they are stable. Among these genotypes, only genotypes G–3, G–10 and G–16 are average responsive, hence these genotypes are suited to all environments, as they produced higher seed yield than population mean, non significant bi. Among the others, the G–12 is highly responsive (significant b1 >1) and higher seed yield than population mean, therefore, G–12 is recommended for favouarable environments.

Though the genotypes G–1, G–4, G–6, G–7, G–8, G–9, G–13, G–14 and G–15 are low or responsive and low seed yield than population mean therefore they are not suitable for general cultivation however they have genes for stability and may be used in crossing programme. Whereas, the genotypes,G–2, G–5 and G–11 have high mean performance but significant S^2d hence their performance in given environments could not be predicted and thus considered as unstable.

From these results of stability, it may be concluded that the genotypes G–3, G–10 and G–16 may be grown safely for obtaining more economical yield potential without facing any major drawback of adaptation to these particular environments.

Additive Main Effects and Multiplicative Interaction (AMMI) Model

AMMI analysis, the flexible set of models proposed by Kempton (1984), Cross (1990) and Gauch (1990) combines ANOVA and Principle Component Analysis (PCA) into single analysis with both additive and multiplicative statistical model. AMMI model helps to quantify the interactions by indentifying the underlying causal physiological and environmental patterns. AMMI is applicable when data are quantitative rather than categorical, when data is structured in a two way factorial table, replicated or not. The data with significant main and interaction effects fit well for the AMMI analysis. AMMI analysis provides the information on 1) quantification of G × E interactions 2) details on environmental factors accounting for G × E interactions and 3) information on stability and selection potential for the genotypes.

Salient features of AMMI Model

(*i*) AMMI first calculates genotypes and environment additive (main) effects using analysis of variance (ANOVA) and then analyzes the residual from this model (the interaction) using principal component analysis (PCA).

(*ii*) ANOVA [Additive model) computes a genotype deviation [difference from the grand mean]and environment deviation whose sum (plus the grand mean) estimates the yield for that genotype in that environment. Example : If the grand mean=100, Genotype deviation=5, Environment deviation=7, then yield estimate is 100+5+7=112.

(*iii*) PCA is a multiplicative model as it computes a genotype score and environment score whose product estimates the yield for that genotype in that environment. Example : 100 (5 x7) =135. More than one PCA axis may be calculated.

(*iv*) It is a hybrid model, the results is the least square analysis, which further graphical representation of numerical results (biplot analysis) and often allow a straight forward interpretation of the underlying causes of G × E interaction.

(*v*) This model is effective in identifying the genotypes that have specific adaptation (interacting) and those which are adaptable (non–interacting). It is also useful for characterizing the environments/locations which are suitable for growing a specific or group of the hybrids.

The AMMI model is :

$$Yij = \mu + gi + ej + \sum hk\ \acute{\alpha}ij\tau\ ik + Rij$$

where,

Yij is the yield of i th genotype in the j th environment ,

gi is the mean of the ith genotype as a deviation from the grand mean μ ;

ej is the mean of the j th environment minis the grand mean (μ) ,

hk is the eigen value of the PCA axis k,

$\acute{\alpha}$ij and τik are the principal component scores for k of the i th genotype and the j th environment respectively and Rij is the residual.

The GE interaction sum of squares was subdivided into PCA axis , where axis k is regarded as having t + s–1–2k degrees of freedom and t and s are the number of the genotypes and environments respectively.

Steps involved in AMMI analysis

AMMI analysis has two parts: ANOVA and PCA. The fist of AMMI analysis was ANOVA to partition the total variation into three orthogonal source genotypes (G), environments (E) and genotype–environment interactions (GE), PCA, second part of AMMI analysis partition the GE interaction into several orthogonal IPC axes (Table). Following are the major steps involved in AMMI analysis.

(*i*) Environment–wise analysis and pooled analysis of variance are conducted as per the normal procedures. If genotypes, environments and G × E interactions are significant, the analysis may be procedded further for AMMI analysis.

(*ii*) Computation of the grand mean.

(*iii*) Computation of the environmental means and deviations (environment effect)

(*iv*) Computation of the genotypic means and deviations (genotypes effects)

(*v*) PCA on the genotype variance and covariance matrix.

(*vi*) PCA on the environment variance and covariance matrix.

(*vii*) Multiplying the eigen values by the degrees of freedom of their respect variance–covariance matrix to get the common eigen values.

(*viii*) Singular value decomposition(L) by getting the square root of the common eigen values

(*ix*) Computation of the genotype scores IPCAI, IPCA2, IPCA3 etc by multiplying the eigen vectors (PCAs) by the corresponding square root of 'l'

(*x*) Computation compute the environment scores IPCAI, IPCA2, IPCA3 etc by multiplying the eigen vectors (PCAs) by the corresponding.

MATHEMATICAL BASIS FOR AMMI ANALYSIS

Additive Model:

Two factor data can be represented as matrix with column representing the different levels of one factors, and rows representing the different levels of the second factor. When an experiment is replicated, each replicate can be represented as a separate matrix. The first step in an analysis of a replicated data set is to determined the means of each cell across all replicates. The second step is to calculate the error variance. Thus an additive analysis of the data matrix would have the following mathematical representation.

$$Y_{ij} = \mu + \alpha_i + \beta_j + P_{ij}$$

Where,

Y_{ij} is the actual value

μ is the grand mean

α_i is the deviation of the mean of the genotype from the grand mean.

β_j is the deviation of the mean of the environment from the grand mean

P_{ij} is the residual deviation not explained by μ, α_i β_j

Singular Value Decomposition

Two way data tables are matrices and these matrices can be evaluated using matrix algebra procedures. One matrix algebra procedure which is effective in characterizing a data matrix is called singular value decomposition (SVD). A mathematical representation of a complete SVD of a data matrix is presented in the following equation.

$$Y_{ij} = \sum_{i=1}^{N} \lambda\, y_i\, \delta_j$$

Where,

Y_{ij} is the smaller number of genotype or environment

λ is the singular value (square root of the eigen value which in turn, is the sums of squares divided by the number of replications.

y_i is the given value for genotype is axis j

δ_j is the eigen value for environmental in axis i

The process of least squares SVD can be best visualized by assuming that a data set is multi dimensional cloud of points (number of dimensions is equal to N). SVD then determines the least squares line through the cloud of points. This axis is 1. A least square line perpendicular to the previous line is then drawn through the cloud of points and becomes axis 2. This process is repeated for N axes with each succeeding axis being perpendicular to all the previous axes. This partitions all the data into the complete set of axes. This operation removes the largest singular value first and then the second largest, etc until all the data has been accounted for. The eigen vectors are the proportion of that singular value which is due to a given genotype or an environment.

Principal Component Analysis (PCA): If prior to applying SVD to a data matrix, the grand mean is removed the mathematical representation for PCA is obtained.

$$Y_{ij} = \sum_{i=1}^{N} \lambda\, y_i \delta_j$$

Thus PCA is nothing other than the application of SVD to a matrix after the removal of the grand mean from each value. This removal results in what is commonly called a covariance matrix. If, as a result of a test of statistical, fewer than N axes are statistically different than noise (error variance) then the last few axes which are not significant can be left in residual matrix, giving the following mathematical representation of PCA.

$$Y_{ij} = \sum_{i=1}^{N} \lambda\, y_i \delta_j + P_{ij}$$

where N is the number of significant axes and is less than the smallest of the number rows or columns and P_{ij} is the residual after removing N axes from the matrix.

Analysis of Variance for Stability–AMMI Model

Source	df	SS	MS	F cal.
Treatments	(ge– 1)			
Genotypes	(g–1)			
Environments	(e–1)			
G*E Interaction	(g–1) (e–1)			
PCA I				
PCA II				
Residual				
Block	(r–1)			
Error	(r–1)(ge–1)			
Total	(ger–1)			

The IPCA SS, MS and scores are computed as in PCA analysis. If ICPA MS are significant and residual MS is non significant, the steps may be continued for development of biplot. The AMMI biplot– I is developed by placing both genotype and environment mean values on X– axis and respective IPCA axis eigen vectors on Y– axis.

Interpretation of the Biplot

From the values of mean and ICPA 1, the genotypes are classified under four distinct classes.

Class I : Genotypes with high mean and positive ICPA I

Class II : Genotypes with high mean and negative ICPA I

Class III : Genotypes with low mean and negative ICPA I

Class IV : Genotypes with low mean and positive ICPA I

Bioplot–II is developed by placing ICPA 1 on X–axis and ICPA 2 on Y–axis. This biplot does not show the additive main effect, but by showing two ICPA axes, it is highly informative on interaction. When ICPA is significant, such graph is useful supplement to usual biplot.

AMMI's stability value (ASV) : was calculated in order to rank genotypes in terms of stability using the formula suggested by Purchase (1997) as shown below:

$$\text{AMMI Stability Value (ASV)} = \sqrt{\left[\frac{\text{SS1PCA1}}{\text{SS1PCA2}}(\text{1PCA1 score})\right]^2 + [\text{1PCA1 score}]^2}$$

Where,

SS = Sum of squares;

IPCA1 = interaction principal component axis 1

IPCA2 = interaction principal component axis 2

In general, an absolute AMMI stability value (ASV) was determined using a procedure that combines IPCA1 and IPCA2.

Identifying adaptable genotypes

(*i*) AMMI model, the genotypes which are characterized by mean greater than the grand mean and the PCA scores nearly zero are considered as generally adaptable to all the environments.

(*ii*) The genotypes with high mean performance and with large value (positive) of IPCA scores are considered as having adaptability to favourable environments.

(*iii*) The genotypes with high mean performance and with large value (negative) of IPCA scores are considered as having adaptability to stress/unfavourable environments.

(*iv*) The genotypes with low mean performance and with large value (positive or negative) of IPCA scores are considered as unstable.

SOLVED EXERCISE

A total of 10 hybrids of Maize was evaluated at three locations (E_1, E_2 and E_3) in RBD with three replications for grain yield/plant (g). The location wise data on grain yield is given below. Worked out the stability parameters for different varieties as per AMMI model.

Grain yield/plant (g) at E_1						Grain yield/plant (g) at E_2				
S.N.	I	II	III	Total	Mean	I	II	III	Total	Mean
1	71.2	76.4	77.7	225.3	75.1	102.6	95.3	94.6	292.5	97.5
2	103.4	98.6	95.9	297.9	99.3	99.9	108.3	98.4	306.6	102.2
3	96.8	88.4	89.3	274.5	91.5	106.8	102.8	112.6	322.2	107.4
4	75.4	82.7	82.8	240.9	80.3	115.0	123.4	118.6	357.0	119.0
5	82.1	84.2	85.7	252.0	84.0	98.6	100.8	108.4	307.8	102.6
6	86.7	91.9	97.4	276.0	92.0	93.3	105.6	102.3	301.2	100.4
7	85.0	92.4	85.4	262.8	87.6	92.6	98.6	89.3	280.5	93.5
8	69.4	61.2	65.0	195.6	65.2	82.4	88.8	89.8	261.0	87.0
9	92.2	88.7	98.4	279.3	93.1	97.6	104.3	92.7	294.6	98.2
10	98.2	106.0	108.4	312.6	104.2	107.7	104.2	113.6	325.5	108.5
Total	860.4	870.5	886.0	2616.9		996.5	1032.1	1020.3	3048.9	

Grain yield/plant (g) at E_3					
S.N.	I	II	III	Total	Mean
1	88.4	96.7	101.4	286.5	95.5
2	106.8	99.2	108.7	314.7	104.9
3	108.2	119.8	117.6	345.6	115.2
4	110.0	105.3	116.2	331.5	110.5
5	118.7	112.2	119.8	350.7	116.9
6	129.7	118.3	125.8	373.8	124.6
7	108.6	108.5	94.3	311.4	103.8
8	100.7	103.8	94.3	298.8	99.6
9	108.2	106.1	98.6	312.9	104.3
10	96.2	92.7	86.8	275.7	91.9
Total	1075.5	1223.7	1063.5	3201.6	

Step I : Analysis of variance for individual location

Sources	df	E_1			E_2			E_3		
		SS	MS	F	SS	MS	F	SS	MS	F
Repl.	2	33.25	16.63	0.89	65.77	32.88	1.35	10.37	5.19	0.15
Varieties	9	3600.48	400.05**	21.49	2081.82	231.31**	9.50	2770.31	307.81**	8.63
Error	18	335.09	18.62		438.11	24.34		641.81	35.66	
Total	29	3968.82	136.86		2585.70	89.16		3422.49	118.02	

Step II: Pooled Analysis

Prepare two way table (Mean over replications at each location)

Variety	E_{1}-	E_2	E_3-	Total	Mean
1	75.10	97.50	95.50	268.10	89.37
2	99.30	102.20	104.90	306.40	102.13
3	91.50	107.40	115.20	314.10	104.70
4	80.30	119.00	110.50	309.80	103.27
5	84.00	102.60	116.90	303.50	101.17
6	92.00	100.40	124.60	317.00	105.67
7	87.60	93.50	103.80	284.90	94.97
8	65.20	87.00	99.60	251.80	83.93
9	93.10	98.20	104.30	295.60	98.53
10	104.20	108.50	91.90	304.60	101.53
Total	872.30	1016.30	1067.20	2955.80	98.5

CF = 2955.80/30 = 291225.12

Table SS = 75.10^2 +----+ 91.90^2)-CF = 4861.30

Genotype SS = (268.10^2+-----+304.60^2) /3 – CF = 1350.36

Env. S.S. = (872.3^2 +---+1067.20^2) /10 – CF = 2043.76

G × E int. = Table SS - Gen. SS - Env. SS = 1467.18

MS due to PE = (Error SS E1 +Error SS E2 + Error SS E3)/ Error df of E1 + E2 +E3

= (335.09 + 438.11 + 641.81)/ 18+18+18 = **26.20**

= 26.20/3 = 8.73

Pooled Analysis of variance

Source	df	SS	MS	F
Envi.	2	2043.76	1021.88*	12.54
Varieties	9	1350.36	150.04*	2.84
Env.x Var.	18	1467.18	81.51*	9.33
Pooled error	54	1415.01	26.20	
		PPE	8.73	

Since the G × E interaction is significant, proceeded further for estimating stability parameters by AMMI model.

Step- III : Analysis of variance for stability (AMMI)

Sum of squares

Trials SS = Total SS from pooled analysis of variance = 4861.30

Genotypes SS = 1350.36

Environment SS = 2043.76

G × E Interaction SS = 1467.18

Total Ss = (Total SS in E1 + E2 + E3)/Rep] + Env. SS

= (3968.82 + 2585.70 + 3422.49)/3

= (9977.01)/3 + 2043.76 = 3325.67 +2043.76 = 5369.43

Error SS = Total SS - trial SS = 5369.43 -4861.30 = 508.13

Analysis of variance for stability (AMMI)

Source	df	SS	MS	F cal.	% of total SS
Trials	29	4861.3	167.63**	19.79	90.54
Genotypes	9	1350.36	150.04**	17.71	27.78
Environments	2	2043.76	1021.88**	120.65	42.04
G*E Interaction	18	1467.18	81.51**	9.62	30.18
PCA I	10	996.55	99.66**	11.77	67.92
PCA II	8	470.63	58.84**	6.95	32.08
Error	60	508.13	8.47		9.46
Total	89	5369.43	-		

The AMMI analysis of variance for grain yield /plant showed significant effects for genotype, environment and GE interaction . These results showed that 27.78, 42.04 and 30.18% of the total trial sum of squares (90.54 %) was attributable to genotype, environment effects and G × E interaction effects, respectively. A higher percent of SS for environments and genotypes indicated that the environments and genotypes were diverse, with large differences among genotypes means causing most of the variation grain yield. Further, first Interaction Principle Component Axis (IPCA1) captured 67.92 % of the G × E interaction SS with 10 df. Similarity the IPCA2 explained a further 32.08% of the GE interaction SS. The mean squares for IPCA1 and IPCA2 were significant and cumulatively contributed to 100% of the total GE interaction. The treatment and block SS combined make up 90.54 % of the total SS, while the error SS make up 9.46 % of the total SS.

AMMI Stability Value is calculated by the formulae

$$ASV = \sqrt{\left[\frac{\text{SSIPCA1}}{\text{SSIPCA 2}}(\text{IPCA1 score})\right]^2 + [\text{IPCA1 score}]^2}$$

Step IV: Estimates of variety-wise IPCA sores and ASV

Genotype	E_1	E_2	E_3	Mean	IPCA 1	IPCA 2	ASV	Rank
1	75.10	97.50	95.50	89.37	-0.06	1.33	1.33	2
2	99.30	102.20	104.90	102.13	1.71	-0.91	2.65	6
3	91.50	107.40	115.20	104.70	-0.53	-0.07	0.77	1

Contd...

Contd...

4	80.30	119.00	110.50	103.27	–1.20	3.40	3.82	9
5	84.00	102.60	116.90	101.17	–1.70	-0.33	2.49	5
6	92.00	100.40	124.60	105.67	–1.74	-2.09	3.28	8
7	87.60	93.50	103.80	94.97	0.36	–1.23	1.34	3
8	65.20	87.00	99.60	83.93	–1.87	0.11	2.72	7
9	93.10	98.20	104.30	98.53	1.00	-0.97	1.75	4
10	104.20	108.50	91.90	101.53	4.04	0.76	5.92	10
Env. Mean	87.23	101.63	106.72					
Env. Index	–11.30	3.10	8.19					
IPCA–1	-3.85	-0.24	4.09					
IPCA-2	-2.07	3.80	–1.73					

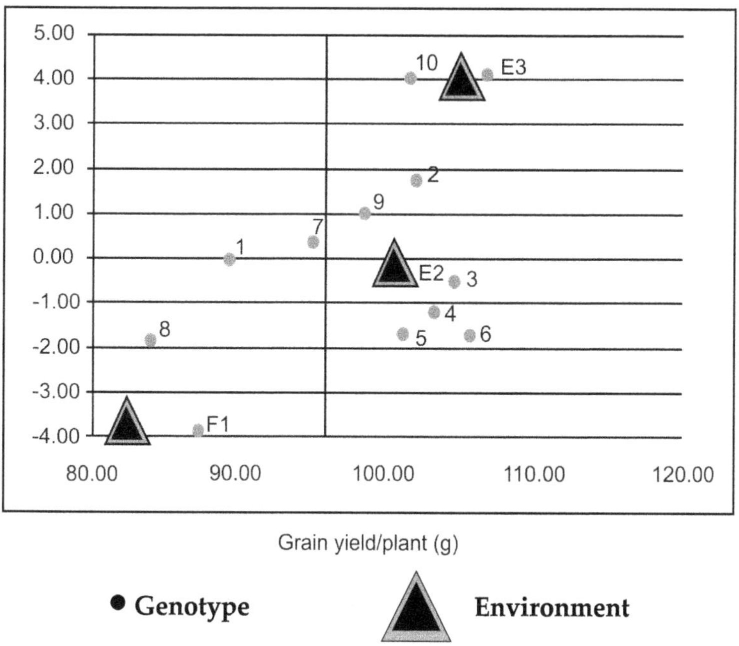

Fig. 1: AMMI- I : Biplot for grain yield/plant (g) of main effects(genotypes and Location) and G × E interaction (IPCA–1) of maize

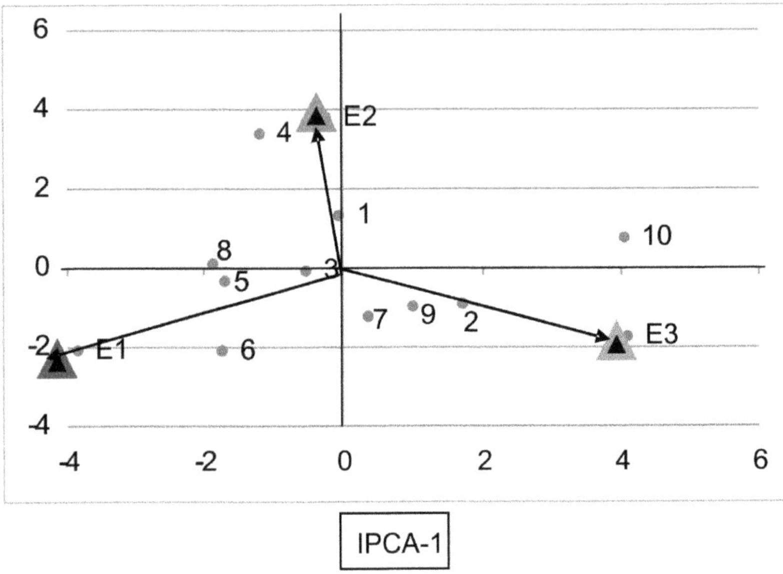

Fig. 2: AMMI- II Biplot for grain yield of G × E interaction (IPCA–1 and IPCA-2) of maize

Interpretation of results

The AMMI I biplot (Fig. 1) showed that the hybrids No.1, No.3 and No.7 had had negligible interaction, indicating stability for grain yield. However, only the hybrid No.3 had high mean than population mean , hence it has average stability. The hybrids No.2 and No.10 had high main effects with high positive interaction, indicating their suitability for favourable environments, whereas, hybrids No.4, and No.5 high main effects with high negative interaction, indicating their suitability for poor environments. Considering the ASV ranking the hybrid 3 was found adaptable for all locations.

Among the three locations, (E_3) followed by (E_2) had high mean but high positive and low negative interaction, respectively. Whereas, (E_1) had low mean with high negative interaction.

AMMI II biplot for grain yield is presented in Fig. 2. The hybrid 3 scattering near the origin, were non-sensitive to environmental interactive forces and possessed broad adaptation to most locations. Judged by the distance of the environments from origin the interactive forces of (E_1) and (E_3) were strong locations, while it was small at (E_2)

Further, from AMMI II biplot, genotypes showing good adaptation for grain yield to specific locations were hybrids No.3 and No.6 for (E_1), hybrids No.1, No.3 and No.4 for (E_2) and hybrids No.1, No.3 and No.4 for (E_3).

34 Transgressive Segregation

Transgressive segregation is defined as the appearance of individuals in segregating populations that fall beyond their one or two parental phenotypes.

Occurrence and Utility of Transgressive Segregation:

It is often observed in the progeny derived from interspecific cross. For the occurrence of the transgression, Rick and Smith (1953) proposed three potential explanations including *de novo* mutation induced by hybridity, complementary action of genes from the two parental species and un masking of recessive genes normally held heterozygote.

The choice of parents is the basic step in hybridization programme. Due to recombination and segregation transgressive segregation are produced in F_2 generation by accumulation of favourable genes from the parents involved in hybrid. Occurrence of transgressive segregants suggest their potential for recovering suitable lines with desirable combination of various traits in crop improvement programme.

It is postulated that when crosses are made between high and low performing parents instead of both high or both low, there is maximum possibility of getting transgressive segregants from F_2 generation onwards since the segregating high genes will be linked or buffered with positive modifies for the accumulation of more positive buffers with more high genes.

Estimation Procedure of TG

The data on individual plants for each character were pooled together and mean, standard deviation, SE of mean, variance and standard variates are obtained as follow

(*i*) **Arithmetic mean :** $\overline{X} = \frac{\Sigma x}{N}$

Where, Σ = summation, × = an observation, N = number of observations

(*ii*) **Standard deviation:** $SD = \sqrt{\dfrac{\Sigma x^2 - \dfrac{(\Sigma x)^2}{N}}{N-1}}$

(*iii*) Variance: $\sigma^2 = \dfrac{\Sigma x^2 - \dfrac{(\Sigma x)^2}{N}}{N-1}$

(iv) **Standard error:** $SE = \frac{SD}{\sqrt{N}}$

(v) **Standard variates** $= \frac{X_i - \overline{X}}{\sigma}$

where, x_i = Variable value for i^{th} individual

$\overline{X}$ = Mean of sample

σ = Standard deviation

(vi) ***Limiting value***

The limiting value of standard variates corresponding to range of parental mean of 5% probability level was calculated so that the segregants beyond this limiting value would be transgrents. Transgressive segregants showing significant deviation only in desirable direction are considered for drawing interferences about transgression.

Threshold value :

$$T.V. = (\overline{P}^{+}) + 1.96\,\sigma\,(\overline{P}^{+})$$

Where, $(\overline{P}^{+})$, and $\sigma\,(\overline{P}^{+})$ are mean and standard deviation of increasing parents, respectively.

The Limiting ND value

$$\text{ND value} = \frac{T.V. - \overline{F}_2}{\sigma \overline{F}_2}$$

Where, T. V. is threshold value & are mean and standard deviation of F_2, respectively.

Graphical Representation

Taking standard deviation as a class interval, graphs of frequency distribution are plotted for character of F_2 generation. The frequencies of individuals in the interval are presented on the vertical axis while the class width taken as one standard deviation is taken on the horizontal axis. The shaped portion in the frequency histogram represents the area covered by desirable transgressive segregants. In each of graph, zero on abscissa corresponds to mean of the respective generations.

Table: Mean-range, SE, SD , frequency of estimated transgressive segregants for grain yield character in F_2 generation of a cross is given below.

Generation	Mean	SE	SD	Frequency of TG							Total	Total TG	%	ND
				-3	-2	–1	0	1	2	3	Plants		TG	Value
P_1	12.02	0.65	3.58	-	-	-	-	-	-	-	30	-		
P_2	14.2	0.71	3.91	-	-	-	-	-	-	-	30	-		
F_2	15.95	0.53	7.51	0	0	0	156	38	6	0	200	44	22.0	0.78

From above data given in table, out of 200 plants in F_2 populations, 44 plants (22 %) were transgressive segregants (in desirable positive direction) and had normal deviation value (0.78) of F_2 population.

References

Allard, R.W. and Bradshaw, A.D., 1964. Implications of genotype-environment interactions in applied plant breeding. *Crop Science* 4: 503-508.

Anderson, E. 1957. A semigraphical method for analysis of complex probems. *Proc. Natn. Aca. Sci. Wash* 43. : 923-927

Arunachalam, V. 1974. The fallacy behind the use of a modified line × tester design. *India J. Genet.*, 33: 280,-287

Becker, H.C., 1981. Correlations among some statistical measures of phenotypic stability. Euphytica 30: 835-840.

Becker, H.C. and Léon, J., 1988. Stability analysis in plant breeding. *Plant Breeding* 101: 1-23.

Bruce,A. B. 1910. The Mendelian theory of heredity and the augmentation of vigour. *Science*, 32 : 627-628.

Burton, G. W. 1952. Quantitative inheritance in grasses. *Proc. 6th Int. Grassland Cong.* 1: 227-283.

Calinski, T., Czajka, S. & Kaczmarek, Z. 1987. A model for the analysis of a series of experiments repeated at several places over a period of years. I. Theory. Biuletyn Oceny Odmian, 17–18: 7-33.

Cavelli, L. L., 1952. An analysis of linkage in quantitative iheritance. In " Quantitative inheritance" Ed. E. C. R. Reeve and C. H. Waddington, HMSO, London.pp. 135–144.

Chahal, G.S. and Gosal, S.S., 2002. Principles and procedures of plant breeding: Biotechnological and Conventional approaches. Narosa Publishing House. New Delhi, India.

Cockeram, C. C. 1963. Estimation of genetic variacnces, pp. 53-94 in Hanson, W.D. and Robinson, H. F. (eds). Statistical Genetics and Plant Breeding. *Nat.Acad. Sci. Nat. Res.* Council Publ. No. 982, Washington, D. C., USA.

Comstock, R.E. and Moll, R.H., 1963. Genotype-environment interactions. In: Statistical Genetics and Plant Breeding. *NAS-NRC.* pp. 164–196.

Combstock, R. E. and Robinson, H. F. 1948. The components of genetic variance in population of biparental progenies and their use in estimating average degree of dominance. *Biometrics*, 4: 254-266.

Combstock, R. E. and Robinson, H. F. 1952. Estimation of average dominance of genes. Heterosis, ed. J. W. Gowen. Ames: Iowa State College Press. pp.494-516.

Cooper, M., Delacy, I.H. 1994. Relationships among analytic methods used to study genotypic variation and genotype-by-environment interaction in plant breeding multienvironment trials. *Theor. Appl.* Genet. 88, 561-572.

Cornelius PL, Seyedsadr M and Crossa J, 1992. Using the shifted multiplicative model to search for "separability" in crop cultivar trials. *Theor. Appl. Genet.* 84:161–172.

Cornellus, P. L., Seyedsadr, M. and Crossa, J. 1992. Using the shifted multiplicative modelto search for "sepearability: in crop cultivar trials. *Theor. Appl. Genet.*

Crossa, J., 1990. Statistical analysis of multilocation trials. *Advances in Agronomy* 44:55-85.

Dabholkar, A.R., 1999. Elements of biometrical genetics. Concept Publishing Company, New Delhi, India.

Devenport, C. B. 1908. Degeneration, albinism and inbreeding. *Science*, 28: 454-455.

Device, R. W. 1927. Report of the plant breeder. Res. Puerto Rice Agric. Expt. Sta. 14: 50pp.

Dewey , O. R. and Lu, K. H. 1959. An correlation and path analysis of components of crested wheatgrass seed production. *J. Agron.*, 57: 515-518.

East, E. M. 1908. Inbreeding in corn. *Corn Agr Exp Sta Rep* (New Haven) 190: 419-428.

East, E M. 1936. Heterosis. *Genetics*, 21: 375-397.

Eberhart, S.A. and Russell, W.A., 1966. Stability parameters for comparing varieties. *Crop Science* 6: 36-40.

Eskridge, K.M. 1990. Selection of stable cultivars using a safety-first rule. *Crop Sci.*, 30: 369-374.

Finlay, K.W. and Wilkinson, G.N., 1963. The analysis of adaptation in a plant breeding programme. *Australian Journal of Agricultural Research* 14:742-754.

Fisher, R.A., 1918. The correlation between relatives on the supposition of Mendelian inheritance. Trans. R. Soc. Edinburgh 52: 399-443.

Fisher, R.A., 1936.The use of multiple measurements in taxonomic problem. *Annals of Eugenesis*. 7: 179:188.

Fisher, R. A., Immer, F. R. and Tedin, O. 1932. The genetical interpretation of statistics of 3rd degree in study of quantitative inheritance. *Genetics*, 17: 107–124.

Francis, T.R. and Kannenburg, L.W., 1978. Yield stability studies in short season maize. I . A descriptive method for grouping genotypes. *Canadian Journal of Plant Sciences* 58: 1029–1034.

Freeman, G.H., 1990. Modern statistical methods for analysing genotype × environment interactions. In: Kang, M.S. (ed.). Genotype-by-environment interaction and plant breeding. Louisiana State University Agricultural Center, Baton Rouge, La. pp. 118–125.

Freeman, G. H. and Dowker, B. D. 1975. The analysis of variation between and within genotypes and environments. *Heredity*, 30: 97:109.

Freeman, G.H. and Perkins, J.M., 1971. Environmental and genotype-environmental components variability. VIII. Relations between genotypes grown in different environments and measures of these environments. *Heredity* 27: 15-23.

Gauch, H.G., 1982a. Noise reduction by eigenvector ordinations. *Ecology* 63: 1634–1649.

Gauch, H.G., 1982b. Multivariate analysis in community ecology. First edition. Cambridge University press, London and New York.

Gauch, H.G., 1988. Model selection and validation for yield trials with interaction. *Biometrics* 44: 705-715.

Griffing B. 1956. Concept of general and specific combining ability in relation to diallel crossing systems. *Aust. J. Biol. Sci.* 9: 463-493.

Hanson W.D. and Johnson H.W. 1957. Methods of calculating and evaluating a general selection index obtained by pooling information from two or more experiments. *Genetics* 42: 421-432.

Hayman B.I. 1954a. The theory end analysis of diallel crosses. *Genetics* 39: 789-809.

Hayman B.I. 1954b. The analysis of variance of diallel tables. *Biometrics*, 10: 235-244.

Hayman B.I. 1958. The separation of epistatic from additive and dominance variation in generation means. *Heredity*, 12: 371-90

Hotelling, H. 1933. Analysis of a complex of statistical variables into principal components. Oxford, England: Warwick & York.

Huehn , V.M. 1979: Beitrage zur erfassung der phanotypischen stabilitat. *EDV Med. Biol*, 10: 112–117.

Hüehn,V. M. 1990a. Nonparametric measures of phenotypic stability: II. Applications. *Euphytica* 47, 195-201.

Hüehn,V.M. 1990. Nonparametric measures of phenotypic stability. Part 1: theory. *Euphytica*, 47: 189–194.

Hull, F. H. 1945. Recurrent selection for combining ability in corn. *J.AM. Soc. Agron.*, 37:134–145.

Jackson, P., Robertson, M., Cooper, M. and Hammer, G.L., 1998. The role of physiological understanding in Plant Breeding: From a breeding perspective. *Field Crops Research* 49: 11-37.

Jinks, J.L. 1954. The analysis of continuous variation in diallel crosses in *Nicotiana rustica* L. varieties. *Genetics*, 39: 767-788.

Jinks, J. L. and Hayman, B. I. 1953. The analysis of diallel crosses. Maize Genetics Coop. Newsletter, 27: 48-54.

Jinks, J. L and Jones, R. M 1958. Estimation of the components of the heterosis. Genetics, 43: 223-234.

Johnson, H.W., Robinson, H.P. and Comstoc, R.E. 1955. Estimation of genetic and environmental variability in soybeans. *Agron. J.*, 47: 314-318.

Johnson, R.A. and Wichern, D. W. 1982. Apllied multivatiate statistical analysis. Prentice-Hall, Inc. New Jersey.

Kang, M.S. 1988. A rank-sum method for selecting high-yielding, stable corn genotypes. *Cereal Res. Comm.*, 16: 113–115.

Kang, M.S. 1991. A modified rank-sum method for selecting high yielding, stable crop genotypes. *Cereal Res. Commun.* 19:361-364.

Kang, M.S., 1993. Simultaneous selection for yield and stability in crop performance trials: Consequences for growers. *Agronomy Journal* 85: 754-757.

Kang, M.S., 1996. Using genotype-by-environment interaction for crop cultivar development. Advances in Agronomy 62: 199-252.

Kearsey, M. J. and Jinks, J. L. 1968. A general method of detecting additive, dominance and epistatis variation for metric traits. I. Theory. *Heredity*, 23: 403-409.

Keeble, F. and Pellow, C. 1910. The mode of inheritance of stature and time of flowering in Peas (*Pisum sativum*). *J. Genetics*, 1:47-56.

Kempthorne, O. 1957. An Introduction to Genetic Statistic. John Willey and Sons. Inc., New York.

Kempthorne, O. and Curnow, R. N. 1961. The partial diallel cross. *Biometrics*, 17 : 229-250.

Kempthorne O. and Nordskog, A. W. 1959. Restricted selection indices. *Biometery*. 15 : 10.

Kempton, R.A. 1984. The use of biplots in interpreting variety by environment interactions. *J. Agric. Sci.* 122, 335-342.

Ketata, H., Smith, E. L. Edwards, L. H. and McNew, R. W. 1976. Detection of epistasis, additive and dominance variation in winter wheat (*Triticum astivum* L. *em Thell*). *Crop Sci.*, 16:1-4.

Lin, C.S. and Binns, M.R., 1988. A superiority measure of cultivar performance for cultivar × location data. *Canadian Journal of Plant Sciences* 68: 193–198.

Lin, C.S., Binns, M.R. and Lefkovitch, L.P., 1986. Stability analysis: where do we stand ? *Crop Science* 26: 894-899.

Lerner, I. M. 1954. Genetic Homeostasis. Oliver and Boyd, Edinburg.

Lush, J.L. 1937. Animal Breeding Plans. Collegiate Press, Inc., Ames.

Lush, J. L. 1940. Intra-sire correlation and regression of offspring in rams as a method of estimating heritability of characters. *Proc. American Soc. Animal Product.*, 33:292-301.

Mahalanobis, P. C. 1928. A statistical study at Chinese head measurements. *J. Asiatic Sco., Dengal.* 25: 301-77.

Mather, K. 1949. Biometrical Genetics. Methuen and Co. Ltd. London.

Mather, K, and Jinks, J L. 1971. Biometrical Genetics, 2nd edition, Chapman and Hall, London.

Nandarajan N, Gunasekaran M. 2005. Quantitative Genetics and Biometrical Techniques in Plant Breeding. Kalyani Publishers.

Nassar, R. and Huehn, M., 1987. Studies in estimation of phenotypic stability: Test of significance for non-parametric measures of phenotypic stability. *Biometrics* 43: 45-53.

Nilsson-Ehle, H. 1909. Kreuzungsuntersuchungen an Hafer and Weizen, Lind.

Perkins, J. .M. 1972. The principal component analysis of genotype environment interactions and the physical measurement of the environment. *Heredity*, 29: 51-70.

Perkins, J.M. and Jinks, J.L., 1968. Environmental and genotype-environmental components of variability. III. Multiple lines and crosses. *Heredity* 23: 339-35.

Perkins, J.M. and Jinks, J.L., 1971. Specificity of interaction of genotypes with contrasting environments. *Heredity* 26: 463-474.

Pearson, K. (1901). "On Lines and Planes of Closest Fit to Systems of Points is Space". Philosophical Magazine Series 6 2 (11): 559–572.

Pinthus, M.J., 1973. Estimate of genotype value: A proposed method. *Euphytica* 22: 121–123.

Plaisted, R.L. and Peterson, L.C., 1959. A technique for evaluating the ability of selections to yield consistently in different locations or seasons. *American Potato Journal* 36: 381- 385.

Purchase, J.L., 1997. Parametric analysis to describe Genotype × Environment interaction and yield stability in winter wheat. Ph.D. Thesis, Department of Agronomy, Faculty of Agriculture, University of the Free State, Bloemfontein, South Africa.

Robinson H.F., Comstock R.E., and Harvey P. 1966, Quantitative genetics in relation to breeding on the centennial of Mendelism. *Indian J. Genet.*, 26: 171–177

Rao, C. R. (1952). Advanced Statistical Methods in Biometrical Research. John Wiley and Sons, New York pp. 390-395.

Rawlings J.O. and Cockerham C.C. (1962a). Triallel Analysis. *Crop Sci.*, 2: 228-231.

Rawlings J.O. and Cockerham C.C. (1962b). Analysis of double cross hybrid populations. *Biometry* 18: 229-244.

Rick, C. M. and Smith, P. G. 1955. Novel variation in Tomatato species hybrid. 88:359-373.

Seif, E., Evans, J.C. and Ballam, L.N. 1979. A multivariate procedure for classifying environments according to their interaction with genotypes. *Aust. J. Agric. Res.*, 30: 1020–1026.

Sivasubramanian, V and Madhavamenon, P. 1973. Path analysis for yield and yield components of rice. *Madras Agric., J.*, 60:1217–1221.

Sprague, G. F. and Tatum, L. A. 1942. General vs. specific combining ability in single cross of corn. *Jour. Amer.Soc. Agron.*, 34: 923-932.

Shull, G.H. 1914. Duplicate genes for capsule form in *Bursa bursa-pastoris*. Zeitschr. f. indukt. Abstamm.- U. Vererb. 12: 97–149.

Shukla, G.K., 1972. Some statistical aspects of partitioning genotype-environmental components of the variability. *Heredity*, 29: 237-245.

Simmonds, N.W., 1962. Variability in crop plants, its use and conservation. *Biol. Rev.* 37: 422- 465.

Simmonds, N.W., 1991. Selection for local adaptation in plant breeding programmes. *Theoretical and Applied Genetics* 82: 363-367.

Singh, R. K. and and Choudhary, B. D.1985. Biometrical Methods in Quantitative Genetic Analysis. Kalyani Publishers, New Delhi.

Singh, Phundan and Nararayan,S.S. 2000. In Biometrical techiniques in Plant Breeding. Kalyani Publications.

Smith, H. E. 1936. A discriminant functionfor plant selection. *Ann. Eugenics*, 7: 240-250.

Snedecor, G.W., and W.G. Cochran. 1967. Statistical methods, 6th edn. Iowa State University, Ames, IA.

Suzuki, D., Griffiths, A. and Lewontin, R., 1981. An introduction to genetic analysis. W. H. Freeman and Company. San Francisco, USA.

Tai, G.C.C., 1971. Genotype stability analysis and its application to potato regional trials. *Crop Science* 11: 184–190.

Tysdal, H. M., Kiesselbach, T. A.. and Westover, H. L. 1942. Alfalfa breeding. Univ. Nebraska Agric. Exp. Sta. Res. Bull., 24.

Weber, W. E. and Wricke, G. 1990. Genotype × environment interaction and its implication in plant breeding In " Genotype by Environment Interaction and Plant Breeding" (ed. Manjit S. Kang) Louisiana State University Agricultural Centre, Baton Rouge, U.S.A.

Westcott, B. 1987. A method of assessing the yield stability of crop genotypes. *Journal of Agricultural Science*, Cambridge, 108: 267-274.

Wricke, G., 1962. Über eine Method zur Erfassung der Ökologischen Streuberite in Feldversuchen. Z. Pflanzenzüchtg 47: 92-96.

Yan, W. and Holland, J.B. 2010.. A heritability–adjusted GGE biplot for test environment evaluation. *Euphytica*, 171(3): 355–369.

Zobel, R. W. 1990. A powerful statistical model for understanding genotype by environment interaction, In " Genotype by Environment Interaction and Plant Breeding" (ed. Manjit S. Kang) Louisiana State University Agricultural Centre, Baton Rouge, U.S.A.

www.ingramcontent.com/pod-product-compliance
Ingram Content Group UK Ltd.
Pitfield, Milton Keynes, MK11 3LW, UK
UKHW021030041026
15470UKWH00013B/1142

9 789351 309703